THE MATHEMATICAL EDUCATION TRUST

TEACHERS
OF MATHEMATICS

Some Aspects of Professional Life

Hilary Shuard and Douglas Quadling
(Editors)

Harper & Row, Publishers
London

Cambridge		San Francisco
Hagerstown		Mexico City
Philadelphia		Sao Paulo
New York		Sydney

First published 1980
Harper & Row Ltd
28 Tavistock Street
London WC2E 7PN

British Library Cataloguing in Publication Data
Teachers of mathematics.
  1. Mathematics teachers – In-service training – Great Britain
  I. Shuard, Hilary   II. Quadling, Douglas
  510′.7   QA14.G7

ISBN 0–06–318174–6
      0–06–318175–4 Pbk

Typeset by Inforum Ltd, Portsmouth
Printed and bound by A. Wheaton & Co Ltd, Exeter

CONTENTS

PREFACE

In 1976, the Third International Congress on Mathematical Education was held at Karlsruhe, West Germany. It was attended by 1600 mathematics educators from all over the world. Among the British contributions was a *Report on the Training and Professional Life of Teachers of Mathematics*. This document was received with interest by the teacher training section of the Congress, and was later published in England by the Royal Society, in a revised version. It was very widely distributed in Great Britain, and aroused much discussion. In Scotland, the Scottish Mathematical Council in 1979 prepared its own version, *The Training and Professional Life of Teachers of Mathematics in Scotland*, substantially based on the Royal Society document.

The Fourth International Congress on Mathematical Education takes place at Berkeley, California, in August 1980. The group of contributors to the 1976 report decided to expand and update their work, as a contribution to the Berkeley Congress. This book is the result. It is intended as a portrait of professional life, in many of its facets. Unlike the 1976 report, it is not a committee document, but an amalgam of the very varied experiences and views of many teachers of mathematics. The experiences of individuals are

never typical, but perhaps a collection of them may show some of the
strengths and weaknesses of the totality of teachers of mathematics and of
their professional life. Portraits may be flattering or unflattering – this is a
collection of self-portraits. If it shows some of the needs for an enhanced
quality of professional life, and indicates opportunities which are available
in some areas of the country and not others, it will have done its work.

The editors and contributors hope that it will encourage many teachers of
mathematics to look at their own professional life, and many LEAs,
teachers' centres, and higher education institutions to consider how far
their support for professional life is adequate and appropriate.

The editors are most grateful to all the contributors for the speed and
enthusiasm of their help. The Mathematical Education Trust has gener-
ously sponsored the book, enabling a group of the contributors to meet and
to plan the work.

CHAPTER 1

THE PROFESSIONAL LIFE OF
TEACHERS OF MATHEMATICS

The idea of professional life

> Good teachers of mathematics are among the most valuable resources of an
> educational system.

The report on the *Training and Professional Life of Teachers of Mathematics*
presented at the Karlsruhe Congress in 1976[1] began with the above words.
In 1980, as in 1976, there are not enough good teachers of mathematics in
Britain to meet all the demands which the educational system places upon
them. The pressures on teachers of mathematics from all quarters are
enormous. Society is evidently relying on them to produce future genera-
tions who are as numerate as they are literate, and who are able to apply their
mathematical knowledge in new technological contexts as well as on the
workshop floor and in the marketplace. These pressures are comparatively
new ones, and indeed the word *numeracy* itself is only twenty-one years old;
it was coined in the Crowther Report on the education of sixth-formers to
give a title to the scientific and mathematical studies which were to balance
the curriculum of the arts sixth-former, just as the science sixth-former, on
the other hand, was expected to undertake studies in the humanities which

would promote his literacy at a high level. Only in recent years has the word 'numeracy' changed its meaning to express the mathematical expectations which society has of all its young people. In order to adapt to these pressures, and to the other demands of their work, teachers of mathematics are becoming much more conscious of the need to strengthen their professional life in all its many facets.

What does the term 'professional life' mean for teachers of mathematics? In this book we shall define professional life to mean all those activities in which a teacher takes part when he is not actually working with a group of children, and which enable him to develop his thinking and his skills as a teacher of mathematics.

In the 1937 edition of the Board of Education's *Handbook of Suggestions for Teachers*, teachers were encouraged to think of themselves as independent professionals:

> The only uniformity of practice that the Board of Education desire to see in the teaching of Public Elementary Schools is that each teacher shall think for himself, and work out for himself such methods of teaching as may use his powers to the best advantage and be best supplied to the needs and conditions of his school. . . . But freedom implies a corresponding responsibility in its use.[2]

When those words were written, it was perhaps more possible than it is today for a teacher to think for himself in isolation, for he operated within social and educational traditions which were changing comparatively slowly. Today, whether a teacher works in a primary, middle, or secondary school, he needs to see himself as one of a team of professionals, each of whom has his own part to play in the general and mathematical education of the children who pass through his hands. His professional life as a teacher of mathematics is directed towards that end. This professional life starts on his first entry into teaching, and continues throughout his career. It includes all the interactions with his colleagues from which he receives help and support, and the exchange of day-to-day ideas about mathematics and its teaching with those with whom he works. It includes all the informal support which a teacher gives to his less-experienced colleagues, and indeed to his more-experienced colleagues, through discussions about curriculum, teaching, and mathematics which take place in a lively staff room or mathematics department. It includes professional contacts with HMI*, teachers in other schools, LEA* advisers, lecturers in further and higher education, parents, employers, and the community at large, in so far as

* Acronyms and other items which are marked * are explained in the Glossary (page 142).

these are concerned with mathematics. It also includes taking part in formal in-service education and training, in reading about mathematics and its teaching, in doing mathematics for its own sake, in writing workcards and worksheets, in taking part in the preparation of guidelines for mathematics or in informal or formal curriculum development projects. In short, it includes all those activities which support a teacher of mathematics in thinking more deeply about his work, keeping up-to-date with today's needs, and teaching mathematics more effectively and with more enjoyment at whatever level he works.

Professional commitment

This book has been constructed from the contributions of many teachers of mathematics in primary and secondary schools, headteachers, LEA advisers, wardens of teachers' centres, and lecturers in further and higher education who contribute to the in-service education of mathematics teachers. Each contributor was asked to compose a 'vignette' consisting of personal writing drawn from his own knowledge and experience, on some aspect of the professional life of mathematics teachers. The book consists of these 'vignettes' interleaved with passages which set the scene and explain the point of view of the contributor.

The book's purpose is to provoke further discussion and debate. Teaching mathematics is a most demanding job. Even in 1973, Ron McLone, in a survey of the employment of graduate mathematicians, found that of those who had entered the teaching profession, only 46% of the women and 59% of the men were still teaching six years later. Among the reasons the others gave for leaving were not only the obvious financial reasons, but

a large proportion were also dissatisfied with the organisation of the profession and the 'sterile' nature of the teaching they were asked to carry out.[3]

This book portrays some of these dissatisfactions, as well as displaying the satisfying aspects of teachers' involvement in their mathematical professional life. Many teachers of mathematics are, at present, discouraged by the fact that society is critical of their present effectiveness, and at the same time is withholding the resources for improvement of the quality of their professional life which would make them more effective. Many of the activities described in this book touch only a minority of teachers of mathematics. For those who participate, the activities greatly strengthen their commitment to the profession, and their enthusiasm to continue to work at developing their teaching and their study and enjoyment of mathematics. Participation leads to commitment. But most participation in professional life is undertaken at the expense of other aspects of life – the family, the garden, service to the community, or holidays. So it should be, but not to the extent which we ask of our teachers at present. Many teachers of mathematics have other priorities as well as mathematics: primary teachers are usually responsible for all the work of a group of children, and only a part of their energy can be given to mathematics, even if mathematics ranks high on their list of priorities. Because of the present shortage of mathematics teachers, many secondary teachers teach some mathematics although it is not their first teaching subject, so they have other more important subject commitments; others who were once full-time mathematics teachers now have prior commitments to administration or pastoral work, to out-of-school activities with pupils, or to other forms of service to the community beyond the school. All these commitments are legitimate ones, but when they are put alongside the development of mathematics teaching, it becomes clear that it is unrealistic to expect the enthusiasm of committed teachers alone to produce a major improvement in the quality of mathematics teaching. Although mathematics teaching takes place in individual classrooms, teachers of mathematics need to be able to spend a greater proportion of their working time in meeting together as members of teams, thinking through their problems and policies, planning and experimenting, if a real increase in quality is hoped for. This book portrays some of the ways in which that time might be spent in developing the teaching of mathematics, and with it, in developing the commitment of teachers to this work.

The context

The book is largely intended for readers who are familiar with the context in which mathematics is taught in England and Wales. However, it was prepared with the added intention of making a contribution to the 1980 International Congress on Mathematical Education, and it is hoped that it will be of interest to an international audience. Problems and difficulties in the teaching of mathematics are not confined to any one country. For example, in 1968, Halls and Humphreys wrote, in *European Curriculum Studies, No. 1: Mathematics*, about the supply of mathematics teachers in Europe.

In view of the counter-attractions for mathematics graduates in other occupations, and the expanding demand for secondary education, it would seem unlikely that the shortage of fully qualified mathematics teachers can be filled for many years to come.[4]

And the following comment on the test used in the National Assessment of Educational Progress (USA) will seem very familiar to readers of the APU* survey.

When respondents understand the pertinent mathematical concepts, they can use the concepts in correct analysis of simple word problems. When they are not completely familiar with the underlying mathematics, or when the problem is more complex, errors in problem analysis are more frequent.[5]

It may well be that the experiences of teachers of mathematics in one country are helpful to teachers in other countries. However, the educational context in which mathematics teaching takes place varies considerably from country to country, and so in the remainder of this chapter we describe briefly the context of the educational system in England and Wales in which teachers of mathematics work, and how this context affects them.

The period of compulsory education in England and Wales starts at age five. Over the last century the school-leaving age, at which compulsory schooling ends, has steadily risen. It was last raised, from fifteen to sixteen, in 1972, and the effects of this rise are still being absorbed in the schools.

The overall pattern of educational provision was set out in the last great Education Act, the 1944 Act, which provided that primary schools should cater for children between five and eleven, and provided, for the first time, secondary education for all, from eleven years of age until at least fifteen. Before the 1944 Act, only a proportion of children had gone on to secondary education, while the rest remained in higher elementary schools until the leaving age of fourteen. The new secondary schools for all were to be of three types: grammar schools providing an academic curriculum for about 20% of the population, secondary modern schools for the less academic, and technical schools for those whose aptitude lay in this direction. In most local education authorities, technical schools never became a reality, and during the 1950s the new secondary school system settled into a selective pattern; those who passed the '11+'* examination received an academic education which for most included the study of mathematics at least up to the O-Level* standard of the General Certificate of Education. Those who were unsuccessful at eleven entered the secondary modern schools*, whose mathematical diet was often largely arithmetical. 'Secondary modern children' were largely taught by non graduate teachers, while those children who were fortunate enough to enter the grammar schools met the many graduate mathematicians who entered teaching in the period just before industry began to be the major employer of new mathematics graduates.

In the 1960s, alongside the world-wide 'modern mathematics' movement, came social pressures which led to the gradual reorganization of the selective system of secondary education into a nonselective system of comprehensive secondary schools. This movement grew gradually across the country, with Labour-controlled LEAs leading the way, until in 1976 a Labour government demanded the completion of the comprehensive reorganization of secondary schooling. For the first time, in the new comprehensive schools, highly qualified graduate mathematics teachers, whose pride was in the work of their academic sixth-formers, found themselves responsible for the mathematics of pupils who could barely subtract two-digit numbers, and for whom algebraic symbolism was extremely inappropriate. Mathematics teachers from the erstwhile secondary modern schools had no less a problem in adjusting to the needs of the mathematically more able pupils.

Comprehensive schooling offered a new opportunity to delay a decision to a later age than eleven as to whether a pupil was able enough to profit from an academic course in the secondary school, but this decision could only be delayed if a pupil was not set on an academic or nonacademic track on entry to the secondary school. Consequently, many comprehensive schools started to place their pupils in mixed-ability groups at eleven, and to delay decisions about streaming until twelve or thirteen or even later. As a result, mathematics teachers who were accustomed to teaching algebra and geometry to able students, and teachers who were used to teaching arithmetic to average and below-average pupils, both found themselves facing classes containing the complete ability-range. Problems of adjustment were probably more severe in mathematics than in most other subjects.

At the same time, 'modern mathematics' became a reality in secondary schools. In England and Wales, the curriculum is a matter for individual decision in each school, rather than being centrally determined. Competing 'modern mathematics' projects, such as the School Mathematics Project (SMP) and the Midlands Mathematics Experiment (MME), drew themselves to the attention of schools. SMP was the most successful of these projects, and in response to the needs of the times, SMP Books 1–5, intended for academically able children and published in the mid-1960s, were succeeded by SMP Books A–H, a less demanding course aimed at the middle of the ability-range, but based on 'modern' content. The majority of secondary teachers of mathematics were adapting their teaching to new content and emphases at the same time that they were adapting their teaching styles to the ability-range of the comprehensive schools.

The raising of the school-leaving age to sixteen in 1972 meant that schools needed to provide suitable work for the mathematically less able throughout the five years of secondary education. For many, the new Certificate of Secondary Education (CSE)* provided a target. This 16+ examination was intended for the middle of the ability-range, down to the 60th percentile, and complemented the academic O-Levels of the General Certificate of Education, which is taken by roughly the top 20% of the population. In practice, CSE mathematics is taken in many schools down to the 80th percentile, or lower, of the 16+ age-group. It is a regionally based, teacher-controlled examination.

The idea that primary education (from five to eleven) and secondary education (from eleven to sixteen or eighteen) take place in primary schools* and secondary schools*, with transfer at 11+, is an oversimplification. Some local education authorities made the best use of existing school buildings under comprehensive reorganization by arranging ages of transfer other than eleven. A quite common pattern is that of a first school (five–nine), middle school (nine–thirteen), and upper school (thirteen–eighteen). In some areas, the age-range of middle schools is eight–twelve. In other areas, pupils leave the comprehensive secondary school at sixteen, and transfer to a sixth-form college for the two years from sixteen to eighteen. Primary schooling may include a transfer from infant school (five–seven) to junior school (seven–eleven). At every age between seven and fourteen, children somewhere in England are transferring to a new school.

However, whatever the large-scale organization of primary schooling, the usual internal organization of primary schools is based on the idea of the class teacher. Each pupil is taught by one teacher for a year, and then usually moves to another class and is taught by another teacher the next year. Each teacher is responsible for all the work of her class during the year, although the school may be able to ensure that specialist subjects such as music are taught by a teacher with a particular talent. Mathematics is almost invariably taught by the class teacher. 'Modern mathematics' and comprehensive reorganization have affected the primary class teacher, too, but their impacts have been different from those felt in secondary schools.

Grammar schools were based on the selection process of the '11+ examination', which included mechanical and problem arithmetic. The primary school arithmetic syllabus was based upon the need to ensure that as many children as possible 'passed the 11+'. With the disappearance of the '11+', external control of the primary school syllabus was removed, and schools became free to respond to their perception of the children's mathematical needs. 'Modern mathematics' was felt in the primary schools to be a broadening of the curriculum from the narrow mental, mechanical, and problem arithmetic to include ideas of shape, sets, and pictorial representation. Practical work became much more prominent, and the phrase 'I do and I understand' was the inspiration of the Nuffield Primary Mathematics Project. Unfortunately, not all primary teachers had the personal mathematical background to take advantage of the new freedom, and in some schools the previous emphasis on rote learning of mechanical arithmetic was succeeded by unstructured exploration rather than an increase in understanding. However, on balance the gains have outweighed the losses, and it is certainly true that more primary children enjoy mathematics, and enjoy a broader mathematical diet with some understanding, than was the case twenty years ago.

The last seven years have been a time of backlash and criticism of education, ranging from the Black Papers to the Prime Minister's speech at Ruskin College, Oxford, in October 1976, which inaugurated a 'great debate' on the quality of education. In 1977 the Expenditure Committee of the House of Commons decided:

> to examine the validity and substance of criticisms from many quarters, and especially from employers, about an apparent deficiency in basic skills among school leavers applying for jobs.

Among the recommendations was:

> We recommend that the Secretary of State should set up an enquiry into the teaching of mathematics.[6]

This recommendation was accepted, and in 1978 the Cockcroft Committee of Inquiry came into being, with the terms of reference:

> To consider the teaching of mathematics in primary and secondary schools in England and Wales, with particular regard to the mathematics required in further education, employment and adult life generally, and to make recommendations.

The Cockcroft Committee has not yet reported, but it is against a background of public scrutiny of the teaching of mathematics that the teachers who have contributed to this book have written. The professional life of mathematics teachers can no longer be a personal matter; it is a concern to all who care about the quality of education.

CHAPTER 2

TEACHERS AS INDIVIDUALS

We start our set of vignettes where every teacher starts – as an individual with his own background, needs, and ideas. The time of entry to the profession is a time of expectation, hope, and trauma for every new teacher who has been preparing for several years for the time when he will have his own class or classes, and will be responsible for how they learn mathematics. The induction of these new teachers into the profession may be carefully planned by an imaginative head of department or headteacher. More often, new teachers are left very much to their own resources to find their own way around the school's organization, syllabuses, apparatus, textbooks, and worksheets. In 1972, the James Report on *Teacher Education and Training*[7] proposed an organized induction year for new teachers, who would work under the guidance of experienced teacher-tutors, and would have a lighter teaching load and opportunities for further professional study. In most local education authorities, shortage of resources has meant that little has been done to implement these proposals, and our first set of vignettes shows some of the realities of entry into the profession of mathematics teaching.

Entry into the profession

In 1979, the Teaching Committee of the Mathematical Association set up a working party to produce a handbook which would provide some guidance for new teachers of mathematics and those responsible for them. In order to collect material for a section of this handbook, Margaret Brown and John Hayter have been monitoring the progress and attitudes of some new teachers of mathematics during the year 1979–1980. These ten teachers completed either Postgraduate Certificate in Education or Bachelor of Education courses in 1979, and started work as specialist teachers of mathematics in secondary schools in September 1979. At intervals during their first year of teaching, they wrote *How I Feel about My Teaching Job*. The extracts below are from pieces of writing done at three different periods: just before the beginning of the autumn term, after one or two days of teaching, and after a term of teaching.

The night before . . .

I have visited my school several times already, and so feel familiar with the staff room and layout of the school. Most of the staff are very helpful and friendly. I am not too worried about the actual knowledge involved in teaching maths because the head of department at my school has been excellent during the summer and given me a great deal of information about names, abilities, books, topics to begin with, etc., so I feel quite prepared for what to teach subject-wise. I think that contact with the school before actually starting is very helpful, and so is having confidence in your head of department. I feel very relaxed knowing that anything I need for my lessons will be in my cupboard when I arrive at school on Tuesday.

I think my greatest anxiety about Tuesday morning is the thought of being a form tutor and having twenty-eight new pupils in the school who are more or less totally dependent on me for information – if I am unable to provide this information, their anxiety will be increased and I shall feel inadequate. I am worried about how my classes will respond to me – whether they will be well behaved or whether I shall have to spend most of my time acting the disciplinarian. I am apprehensive about the fact that this is my first job and whether I can cope with it and whether I shall like teaching. I also have some guilt feelings about the fact that I am being paid to teach a subject that I don't consider myself particularly good at, and feel I am conning the pupils in some way. However, I think my greatest worry about Tuesday morning is arriving at school on time – ten to nine seems very early!

And twenty-four hours later . . .

Why I ever decided to teach this year is the question I've been asking myself since I finally sat down for the first time today at 5.30 pm, having missed both breaks today and having been far too busy to eat lunch.

This morning I felt quite confident that I knew what I had to do with my tutor group from ten to nine until noon. However, when I actually led them from the main hall at 9.15 am to go to their form room, I had a horrible feeling of panic, fear, and being totally alone. The only consolation was that they looked ten times more frightened and bewildered than I did (I think!). I forgot to let them out for morning break at 10.30, and when I finally realized, there was only five minutes left – this, I felt, didn't create a very good impression with them.

Lessons began after lunch with a small group of pupils making up the bottom band third years – my head of department had assured me that some of them could be very difficult, so I wasn't feeling too well at this point. However, my previous experiences of behaviour problems and what this school considers behaviour problems seemed to differ considerably, if this lesson was anything to go by. The class was noisy but quite pleasant and cooperative.

By the end of the day I was feeling very tired, totally confused by all the administrative work I had to get through with my new form, and very frightened by what I knew I had to complete in the near future. Administration seemed to occupy my mind more than teaching maths, and the school which earlier this summer seemed so clearly set out and logical suddenly became a confusing mess of pupils, rooms which all looked alike, and a maze of paths all leading to the wrong building.

The knowledge that even now at 7 pm I haven't finished for the day seems very depressing. I'm far too tired to eat and I'm unable to decide whether to prepare lessons for tomorrow or do administration work for my form. The only consolation is that the pace cannot get worse. Despite being totally exhausted, I think I enjoyed my first day of work. I was pleasantly surprised by the cooperation of the pupils, and the patience of my form, who had me to put up with on their first day in a new school.

Before they started, none of the ten new teachers were unduly worried about the actual teaching of mathematics . . .

My feelings have been mixed all day, alternating between fairly fatalistic and absolutely terrified! I think I'm vastly overprepared for tomorrow's lessons (all two of them) and that I'll spend most of the time finding out names and handing out books.

*   *   *

I don't suppose I could be more unprepared – not anyone's fault in particular; teething troubles of two schools merging and a new head of department organizing 'order out of chaos' don't help. I thought it would help to go into school the day before anything happened – so did Alan and Frances from the maths department, but no one else appeared and, as we were all new, we were of no help to one another, except that I found some notes suggesting what books I might use with some classes.

*   *   *

Having worked on the mathematics which I shall be teaching, I feel fairly happy about this, but less prepared for my two periods of chemistry teaching. I went to school today to prepare Friday's chemistry lesson and to brief myself from the notice board. In fact there were only two other teachers there.

*   *   *

The fortnight I spent at the school at the end of the summer term reassured me that the girls there are much better behaved than the boys at my teaching practice school.

One of the classes I'm most anxious about is the everyday maths fourth-year – the school has no books suitable for this. The maths centre is sending me a list of what books might be suitable, but even if I can persuade my head of department to order them it will probably be at least a month before they arrive.

*   *   *

Even though I know the staff and some of the pupils, having completed a long teaching practice at the school, there are still so many unknowns. These are mostly to do with the administration required at the beginning of the year and in running a tutor group. There are so many little details I don't know about, dinner money, books, class lists, etc. I would have liked a day at school today, getting familiarized with all these things, instead of just having a staff meeting tonight.

*   *   *

Everything seemed okay until the staff meeting this afternoon, when I was hopelessly lost moving from one room to another – that was with everyone looking after me, so what will it be like when it's full of kids and I have to fend for myself? There's so much to remember just for the day's timetable, form periods, assemblies, registrations – thank goodness I haven't got a form.

*   *   *

I wish I was one of the 'unlucky' ones who haven't got a job. I'm not worried about the actual teaching part of the job. It's the relations with the rest of the staff. Do I call them Mr. or Mrs. or call them by their first names? Will the clothes I've got be suitable or will there be objections to what I want to wear? Will the other staff treat me like a pupil because I'm new, straight out of college, and can only afford a bike to get about on?

*   *   *

After the first few days . . .

Happy! I've enjoyed my first two days very much. I did eventually sort out all the administrative problems. It certainly helped to be giving out books for the first lesson rather than having to teach. It helped me to start forming a relationship with classes in a more informal way. But having said that it's also good to have the first lesson over with and get settled into a routine.

How do I learn the names of 200 pupils plus my tutorial group?

*   *   *

That was awful, one day with only two lessons, and I feel ready to chuck it in! The first-form first, and they were super – responsive, interested, and well behaved. Then the fifth, a low CSE group. They were rude, disobedient, and generally got me down. Surprisingly, they did work when I really got onto them, but that was my first lesson – I'll have lost my voice by the end of the week. I was just so glad when the hour was up, but it's left me with a horrid sinking feeling and doubtful if I can face tomorrow. Still at least I don't have them, there can't be others as bad!

*   *   *

The school has taken 1½ weeks to settle into the routine of timetabling-as-normal, which has meant that on many occasions up until now I have had classes I have not been able to teach as they have not been fully sorted out. Mathematical puzzles I love you!

When I hear myself shouting in class (to keep the noise down) I feel I have seen it all somewhere before and it quite horrifies me to see myself slipping so easily into what could be a predictable and somewhat boring teacher. I propose to do something about this once I feel I am on top of the work, which hopefully will take weeks rather than months.

I shan't be handing in my notice just yet!

*   *   *

On the whole, things are going fairly well; nice school, colleagues, and (for the most part) pupils. Major problems: discipline (much better, though, than at my teaching practice school), absence of maths textbooks – (the head of department likes to work almost exclusively from worksheets – I think there must be some areas where a textbook can rival anything the maths department of our school can cook up over break on a Banda sheet. Worksheets cause a lot of mess and I suspect the girls sense that material is provided on a rather ad hoc basis, and that there is little overall direction to their course).

I was surprised how soon they started to play up – the honeymoon period lasted about two days. I feel as if I'm now in for several weeks trench warfare – where I and the girls struggle to gain psychological ascendancy.

*   *   *

I feel I have established myself as the teacher, unlike all the time I was on teaching practice, when the forms always treated you as someone who was not really the teacher. I was not well prepared by the head of department and soon found out the law of the jungle concerning equipment and books. I will start homework and more defined lessons instead of working straight from the book. The only trouble I am having at the moment is with a few fifth-formers, but yesterday I played football with most of them and this had some effect in breaking down hostilities. All the forms work well but lack ability. I even have a sixth-form CSE retake class.

*   *   *

As yet my worries are about their mathematical progress rather than their behaviour in my classes, though I daresay that will come. Much to my amazement my third-form (bottom set) did not object to sitting in alphabetical order and one to a double table, and the second-year work in an awesome silence!

The fourths are eager, but there are so many of them. As I talked, the silence 'got' me – did they not understand, had I made a mistake, or are they good? What a task to assess what each is understanding and whether they are working to capacity.

The sixths – nine of them – some doing further maths, some not, some with physics O-Level, some not. I didn't think that proving $v=u+at$ would be hard, but when their O-Level maths hasn't covered much differentiation, never mind integration, it's no wonder they got lost. That was last lesson Friday – I looked to see when we met again – first two lessons Monday. So I spent some time thinking out how to reinstate their confidence, which I seemed to have shattered.

*   *   *

I find lower sixth A-Level seems a lot easier now than it ever did for me, but as my sixth-form are rather bright I am finding it almost impossible to keep one step ahead, which I hate.

I am thankful that I have two third-year and two fourth-year classes of pretty much the same level, as it is only two-thirds of the work for each class – it never quite goes as low as half the work as the classes always seem to tackle things in different ways.

At Christmas, the new teachers were able to look back on their initiation, most of them with some satisfaction at their survival . . .

The main feeling at the end of the autumn term was one of weariness, to the extent that for the last two weeks of term the thought of not being in school for nearly three weeks became dominant. Christmas end-of-term activities caused a slackening of routine, but there were end-of-term examinations, the results of which would appear in reports to be completed early next term. When I first started testing I was dismayed at the results. Was I teaching or were they learning anything? By the time examinations were marked I could accept the results, which seemed better than the tests anyway. I am conscious all the time that it is difficult to inspire children day after day. The most troublesome are revealed soon, then later the most able. These take up most attention and effort, and I am left feeling that the quiet, average-ability children get less personal attention.

*   *   *

I've survived, but how well? Only time will tell. I feel as though I have so many problems to overcome before I succeed, but then who measures success? It's hard when you're the sole judge of what you're doing. Not the least of the problems is the never-ending struggle of getting through to those who aren't bright, aren't willing, and hence hinder the progress of those who are willing, whether bright or not.

On the whole it's been good – I've survived and enjoyed the term despite more 'downs' than 'ups' – a good job that one 'up' is worth several 'downs'!

*　　*　　*

I feel I have survived the first term, but am feeling very tired and will be glad of a rest. The worst thing was waking up in the mornings and going in all enthusiastic and seeing all the old teachers who have lost their commitment and don't do anything for the kids. Another disheartening thing is having to teach an academic syllabus to children who are quite weak and will never use what I am trying to teach, and so this creates an immediate dislike for the subject. On the other hand you still feel good when you have a class all working hard and seemingly understanding and trying. It is important to get involved in things outside your subject, and any time spent helping them outside the class is often repaid many times over inside the class.

*　　*　　*

Ordinary school work takes up so much time that I wonder how I shall ever manage with a form of my own to look after or with taking on extracurricular activities (the latter an important part of the reason why I decided to take up teaching).

*　　*　　*

I like it! I have thoroughly enjoyed my first term as a teacher. At times it has been frustrating, at times rewarding, and at times amusing. The frustrations have come from the staff and school administration, when they have been inadequate and poorly equipped to deal with the pastoral problems of probationary teachers or the pastoral problems of children. Even as a 'beginner' it is easy to spot the defects. I hope I don't become immune to them in a few years.

It has been very hard work, certainly the most demanding thing I have ever done. Lesson preparation, marking, etc., were much more time-consuming than I had expected and the physical effort required to teach quite a full timetable is quite surprising. I come home every day and have a kip before tea!

But you don't battle alone . . .

Before the beginning of term I went to a meeting of probationers in the LEA and after a good 'team talk' the advice we were given seemed to make more sense than all the first term of Piaget. Instead of all the talk about helping, guiding, and caring, we were given a lecture on 'us' and 'them' and how to survive, which at that time seemed more relevant than all the textbooks I had meant to read.

*   *   *

Advisers from the LEA are very helpful people once the first approach has been made.

*   *   *

I could not have lasted without the support and friendship of the staff, both within and outside the department, especially the head of maths. I should have also found it much more difficult if I had not resolved to get involved with activities outside maths, all of which have proved very rewarding. I still like kids – most of the time – so I think I have good reason to stay. I predict that I shall get better. [A]†

Conversations

The next group of teachers are more experienced. They are class teachers in primary schools, and some of them are responsible for the organization of the mathematics teaching in their schools. They are taking part in an in-service course for primary teachers, and the conversations recorded here are with the course tutor, T. The teachers are A, B, C, D, and E.

A   Every month or two, the local adviser has a meeting with all the people who are responsible for mathematics in the primary schools . . .

T   Were you there the other day?

A   No, the car broke down. But on the last occasion I was very depressed about the people in charge of maths because they were . . . well, two of the worst offenders were saying 'Sit them down and make them learn how to do sums.' I seemed to be a lone voice asking for the understanding of number before one could use it in a mechanical way as a basic skill.

I came up against this another time when I was with an educational psychologist, because she was saying that this particular child would

† The author or authors of each vignette are identified by the letter beside their names in the List of Contributors (page 000).

never have any number sense – she had a complete block, and the only way she would be able to cope in school in mathematics was by learning rote methods for mechanical arithmetic.

T   And this was an intelligent child?

A   Yes . . . so I was furious and I cited ways in which that particular child had shown that she could cope with a variety of problems using a counting board and counters. She was able to do not only addition, but subtraction as well up to 'hundreds, tens, and units' . . . and she understood! ·

T   How old was she?

A   Nine – third-year junior.

T   Why had she been referred to the psychologist?

A   Difficulties at home – and I suppose the trouble had centred on her weakest spot in school.

*   *   *

A   We voted for 'equal additions' some way back, a long time ago, and that was apparently the rote, mechanical method that the child had learnt. She had been taught to disregard common sense; in the remedial group I was trying to inculcate a total number sense, and so these things worked against one another.

*   *   *

A   . . . that was what annoyed me so much about the group . . . because they hadn't been on the course they didn't see it . . . they couldn't appreciate it . . . it was as if we had two completely different philosophies . . . all of us on the course approached it in this way . . . it was as if a lot more had been brought in to the course than purely the fundamental concrete stuff we had done – we had acquired a philosophy.

*   *   *

T   You were saying something about a different idea of mathematics. Where would the original idea have come from?

B   Schooldays – my schooldays.

T   . . . and college? What happened at college?

B   Well, not much really. I remember playing with Dienes' blocks, but that was just a six-week curriculum thing for about an hour.

T   Were you 'first school'* . . . or junior* . . . or?

B No, I trained for secondary teaching, so that was probably why.
T With what subject?
B Drama. So really I had a block about maths . . . and I hated it . . .
   really didn't enjoy teaching it at all . . . and one comes into a junior
   school and sees the Beta books and the same thing that you yourself
   used to do, and the alternatives aren't presented to you . . . you have
   to think of them yourself and you've got this block against them . . .
   the course helped me so much to . . . I suppose to loosen up and enjoy
   it all. I had enough confidence in myself after it to wait until something happened with the children . . . to see patterns in numbers for
   the first time – I got quite excited about that. . . .

                    *   *   *

C . . . as there were two of us on the course from the same school, we
   would discuss it the following day at break time, informally. It isn't
   enough for one of us to go on a course.
T I know. I went into another school – it happened to be in break time –
   and in the staff room the 'number cards' from the course were out on
   the table, and the two people who were on the course, plus the head,
   plus all the other members of staff, were in there . . . and it wasn't a
   formal atmosphere, not a specially called staff meeting . . . they were
   just having coffee, but they were playing and talking so that the ideas
   got over to the other members of staff as well.

                    *   *   *

T Can I ask you what you liked about the course?
D Well really . . . being forced to think about such absolutely basic
   things as what is involved in addition, subtraction, multiplication and
   division. I don't think I had ever thought back to square one and tried
   to work it out. I found it very difficult, and I don't think I could ever
   have come out with any good or complete answers on my own, but
   what came out of the course was very, very helpful. I also think that
   some of the games and activities that were thought up were tremend-
   ously helpful; those were very positive things. I really was very
   clueless about things like number lines and structural apparatus for
   fractions. I've had a student this term, so I've been able to do things
   with a counting board and counters, which have been more valuable
   than anything I've done out of a book. I seem to have done everything
   with the counting board: place value, addition, subtraction, using

games that were suggested on the course – the children adored them. To start with the children revealed a horrifying lack of knowledge; it is quite staggering to do these things and realize how foggy they are.

*   *   *

E   It was very interesting comparing notes with other people, that was a good thing that came out of the course – other people's difficulties, successes, and failures.

I think what was least successful . . . but perhaps you don't want to go into that?

T   Oh yes I do!

E   . . . was when we were in groups and were required to come to some set of results on certain topics and problems. I always felt we wasted a tremendous amount of time and didn't come to anything very valuable. [B]

CHAPTER 3

IN THE PRIMARY SCHOOL

No teacher works alone, and certainly no teacher of mathematics can think of himself as working alone. A child's formal mathematical education takes place over a period of at least eleven years, between the ages of five and sixteen. During this time he will probably be taught mathematics by seven or eight people – if he is fortunate. If he is unfortunate, he may well meet a dozen or more teachers of mathematics, all with their own individual approaches and ideas, all with their own strengths and weaknesses and their different ways of explaining and putting things. In these circumstances, it is essential that the teachers whom a child meets over the years should work together as closely as possible, so that the child can recognize a common language and set of ideas and expectations in his mathematical work, and so that his mathematics can develop steadily over the years.

The team of teachers in a primary school is usually made up of general class teachers, many of whom would not claim any special expertise in mathematics. They need to work together to ensure that every child progresses through mathematical experiences which are appropriate to his understanding and his needs. The team in the primary school, like all teams, needs leadership. Traditionally, this role has been occupied by the headteacher, not only in mathematics, but in every area of the curriculum. The first vignette in this chapter is written by a headteacher whose strengths and experience uniquely fit him to carry out this task in mathematics.

A primary headteacher

I am not a typical primary school head since I have been fortunate enough to have had many unique experiences in mathematics, including involvement in the Nuffield Primary Mathematics Project, BBC television, the TAMS Feasibility Study, and at present, the Assessment of Performance Unit. My school is also a very large junior mixed and infants school, with a roll of five- to eleven-year-olds which has fallen from 938 in 1974 to 835 in July 1979, because of the declining birth-rate. There are at present twenty-seven teachers and myself on the staff. As a school, we are involved in several national activities – we took part as a pilot school in the HMI Primary Survey[8], and we play host to two NORMAC (Northern Mathematics Council) courses each year, as I am chairman of NORMAC. These two courses are organized by one of our two deputy heads, while the other deputy and some of my senior teachers greatly assist the operation, which provides courses for some 360 teachers each year. However, these involvements enable me and my colleagues to meet many teachers and to see a much wider field than our own school. I think the problems everywhere are much the same, and the answers are not simple ones.

Whatever else we say or propose, we have to rely on teachers every time. There are now more schemes, textbooks, and materials on the subject of primary mathematics than ever before, but these are ineffective unless handled by a competent and effective teacher. It could also be said that there is too much advice around for comfort, and often we are bewildered by it all.

It is now fifteen years since I was first seconded to the Nuffield Project and was the teacher in the film *I do and I understand*. It was the age of 'discovery' and 'practical work' but it wasn't long before we found out that some children just won't discover anything if left alone without teacher involvement, and that practical work needs much organizing. These modes of teaching are certainly very valuable, but there is no single way; there are many ways and each is part of the whole attempt to generate better mathematics teaching. There is still a place for rote learning, and certainly there is a major place for oral work – talking about mathematics. There are skills and concepts to be considered, and there is the application of these. We say we are now teaching *mathematics*, not just arithmetic, so the subject is now wider and deeper and many teachers find it difficult to cope with this breadth of activity. They then return to the security of the textbook and the 'four rules'.

Obviously, the majority of teachers need fairly regular courses, and in-service training should be a first priority for every education authority. Attending courses voluntarily attracts the better teachers, but those who need it most are the ones who seldom go. A primary school staff is largely made up of married women with domestic commitments, and the quick course after school is not the best time for them – nor is the residential course of from three to ten days. In fairness to these people, it also helps to encourage attendance if there is careful selection of lecturers and course leaders; we need people who are capable of understanding primary school teachers, have a sympathy, and can also put across their subject in a lively, acceptable way. Too often have I been told that 'it was a waste of time going'.

School 'teach-ins' are, in my opinion, the most valuable form of in-service training. In my own school we try to organize at least one each half-term, mainly concentrating on mathematics and science, but other subjects are also included. These 'other subjects' are, of course, important to the majority of teachers who, unfortunately, may be fearful of and may even dislike mathematics. Though I take the 'teach-ins' in my own school, and in others, it is obvious that many heads feel inadequate to do this and when they have no one else who can do it on their own staff, they need help from a source outside the school.

Young teachers just out of college, new members of staff and long-term supply teachers† can be a worry to any headteacher; indeed, it can also worry teachers when they come to a school where modern approaches to mathematics are given prominence. It all takes time; it seems to be a continuous process and there needs to be a build-up of confidence between the head and certain members of staff; the deputy or teacher responsible for mathematics can often make this link between the head and a teacher who is worried about the school's style. If one wants to be good at anything then time must be spent doing it: practice, reinforcement, and persistence. This applies to teachers as well as to children.

A skeleton scheme of work which outlines the basic programme for teachers in a particular year is essential, but it must allow for flexibility; the 'musts' need to be distinguishable from the 'desirables'. In my own school, year meetings are called for discussion of salient features of the term's work

---

† A supply teacher is a teacher who is temporarily attached to a school in place of an absent member of staff.

and any deficiencies which have been noted from the previous year's work. Progress books are kept by each child in which a piece of work is entered approximately once each week, so that at any time a teacher can make a quick appraisal of a child's progress. In the third and fourth years particularly, a detailed profile on each child is kept by his teachers; this will eventually be forwarded to the child's secondary school.

Inevitably, as mathematics is such an important subject, we are forced to consider at least a degree of specialism, perhaps until we have a supply of more mathematically inclined teachers. I am sure that with teachers of the right calibre we could improve our standards in mathematics, and with adequate in-service training we have many such teachers already. As a priority for improvement, I would look very closely at the possibility of extending top infant* children more. More time needs to be spent doing mathematics and every opportunity must be taken to include purposeful mathematics in other subjects, geography and science in particular.

In discussions with secondary teachers lately, their comments on new intakes of children from primary schools were that far too many could not use an atlas, had no sense of direction, and could not measure with confidence or read calibrations accurately. Primary school teachers may well feel bewildered and claim that we try to do too much. This is possibly true, but we should aim to do as much as we can and make more sensible use of our time, the topics we do, and the little money we have to spend on education. [C]

*   *   *

The case for compulsory in-service training

Some teachers would like to be able to see much more (and more expensive) support for mathematics provided both by the LEA and other agencies outside the school. The next headteacher takes the view that substantial progress in primary mathematics cannot be expected without a large-scale effort on the part of government. He is an advocate of compulsory in-service training in mathematics for all primary teachers.

A primary headteacher

1   *Probationary teachers*: Once a student has successfully emerged from her initial training, and has been appointed to a post as a probationary teacher, her training should continue until her probationary year has

been successfully completed. She will have had three or four years of study, the content of which should have included basic mathematical teaching. When she faces the real situation she will naturally wish to turn for help to familiar support. Her motivation will be very high and she will have specific needs. Her training should continue under the joint supervision of the local college of education, the local authority service, and a specially appointed 'mentor' within the school itself. The in-service training in mathematics provided for this first year should be entirely within the school so that the probationary teacher is applying her training in the real classroom situation and to the age-group she is teaching. She will therefore receive maximum support and benefit from her studies.

2   *In-service training for experienced teachers*: There are three types of mathematical support required here. They are as follows:

a   *Responsibility post*: If the government really wishes to emphasize its concern about the teaching of mathematics within primary schools, then it must be prepared to pay for it. The government should reintroduce the *Scale 3 post* within all primary schools. This special responsibility post could be used to offer financial inducement to a teacher to become a real expert in the teaching of mathematics. She would need to develop a special quality of being able to guide and influence other teachers, lending them all the support they need. Such a person would be obliged through the scale post to attend at least one major course every two years just to keep abreast of the subject. The additional costs resulting from this scheme should be borne by central government as a mark of their real intention to try to halt the deteriorating situation. If these SR (special responsibility) teachers showed special aptitude then they could become local coordinators under LEA advisers for mathematics – again a post that would require reward, for instance at Scale 4.

b   *In-service refresher courses*: Regular in-service courses of the highest possible quality should be a feature of each LEA. Every teacher should be expected to attend a refresher course in mathematics at least every three years and this should be documented. These area courses would be augmented by in-school courses organized by the SR teacher. We are now in a period of rapidly accelerating change and teachers must be prepared to be educated for change. The LEA for its part must offer such courses in term-time and offer to cover the teacher's absence through authorizing supply cover. This would encourage heads and staffs to subscribe fully to such area courses. For fourth-year junior school teachers, such courses could be organized on a secondary feeder-school basis so that some form of agreed curriculum approach could be identified. The teachers of children transferring to the same secondary school would have a chance to work together and work with their secondary colleagues to offer some form of continuity within the system.

The in-service courses for classroom teachers should be practically biased and aimed at the age and ability being taught by that teacher. The courses should aim to build a teacher's confidence in the subject. A workshop approach would allow teachers to prepare material of direct relevance for their own classes.

c *Heads' and deputy heads' courses*: All primary school headteachers and their deputies should be obliged to attend at least one major mathematical course every two years. There should also be a mathematical content in all general heads' courses so that the subject can be fully and openly discussed within a wider context of curricular studies. Such courses should not be squeezed into weekends, but should be an essential part of the school year – to engender the idea of continuing education, especially amongst professional teachers. [D]

Mathematics coordinators

In many schools, the headteacher is more fitted by interest and qualification to carry out a leadership role in other areas of the curriculum. Consequently, many local education authorities have begun to appoint, in primary schools, teachers with special responsibility to coordinate and organize the work in mathematics throughout the school. The recent survey of primary schools undertaken by HMI, *Primary Education in England*,[8] found that

> where teachers who held posts carrying special responsibilities had a strong influence in the school this was very strongly associated with good match for all ages and all abilities.

The 'good match' referred to was between the abilities of the children and the suitability of the work they were given for their abilities. HMI went on:

> These findings suggest that where a teacher with a special responsibility is

knowledgeable and able to give a strong lead in planning and carrying out a programme of work, this is effective in influencing the work of other teachers in the school. This in turn would appear to raise the levels of expectation of what children are capable of doing, particularly in relation to the most able children, who were often the least likely to be given work which would extend them intellectually.

The task for the responsible teacher is a demanding one. HMI describe it in the following way:

> Teachers in posts of special responsibility need to keep up-to-date in their knowledge of their subject; to be familiar with its main concepts, with the sub-divisions of the subject material and how they relate to one another. They have to know enough of available teaching materials and teaching approaches to make and advise on choices that suit local circumstances. And they should be aware of the ways in which children learn and of any sequences of learning that need to be taken into account. Additionally, these teachers should learn how to lead groups of teachers and to help others to teach material which is appropriate to the abilities of the children. They should learn how to establish a programme of work in cooperation with other members of staff and how to judge whether it is being operated successfully. They should also learn how to make the best use of the strengths of teachers of all ages and to help them to develop so that they may take on more responsibility. Particular care should be taken to foster the special qualities of intuitive and gifted teachers.

The next extract is a 'job description' for a teacher with special responsibility for mathematics, prepared by the mathematics adviser in one local education authority. This teacher has the title of mathematics coordinator in this particular authority.

An adviser

1 *The mathematics scheme*
   This should be drawn up by the coordinator in consultation with the headteacher and possibly some members of staff. It should be up-to-date, and meet the needs of both the teaching and learning situations across the full ability-range. The scheme will need revision from time to time, and should allow freedom for teachers to develop their own ideas yet support the teacher whose knowledge of mathematics is inadequate. Links across the curriculum should be made.
2 *Mathematical considerations external to the school*
   The coordinator should be familiar with other schemes and new resource materials. The local teachers' centre and the advisory services

are useful contacts to make. The coordinator should be able to make recommendations to the head for the purchase of new books, cards, and equipment.

3   *Liaison with other schools*

Contact should be established horizontally with other primary schools and vertically with the first school and high school(s) and especially should relate to the ages of transfer (e.g., 8+ and 12+) to ensure proper progression and continuity in mathematics.

4   *Assessment and records*

The coordinator should be able to devise a scheme for continuous assessment so that simple, clear records are kept of each child's progress; the aim is to promote good continuity, so that pupils are neither lost nor bored when they move on to a new teacher or school.

The coordinator will need to know the background, experience, and degree of confidence of the teachers with whom he is working. Through workshops and discussions he should appreciate the different modes of teaching going on in his school, and try to pull together the best elements of each. Meetings may involve the whole staff or just the teachers across one or at the most two year-groups. These meetings should be seen as ones where all teachers have something to contribute, where difficulties can be revealed and suggestions made to resolve them, and where new innovations can be put forward and sometimes adopted. Teachers should be well informed about the resources available to them.

The coordinator should be aware of where and how some teachers need support, especially probationers. He will be expected to give advice and help, and will find it necessary to consult agencies outside the school for some of the answers.

Essentially the coordinator must have, and be seen to have, the full support of the headteacher. It is vital that the first in-service session should be successful, with the head actually taking part. As the confidence of and the confidence in the coordinator builds up, teachers will feel able to turn to him for support and guidance.

The coordinator needs, therefore, to be a good listener, to be able to make clear decisions especially on matters of mathematical policy within the school, and to take steps to ensure that they are carried out. He should be able to illustrate his points by personal example, especially with respect to mathematical knowledge, teaching skills, and good classroom organization. He will need to be aware of the leadership skills of 'tell, sell, join, or consult' and to use them appropriately and sympathetically. [E]

The coordinator's task

But how does it work out in practice? Clearly it is important for the coordinator to organize a good mathematics syllabus within the school and to have an effective recordkeeping system; but in the long term the health of mathematics teaching depends on strengthening the effectiveness of the staff as teachers of mathematics. Yet in an informal survey conducted by a mathematics adviser in another part of the country, while 70% of mathematics coordinators thought that they *should* 'run in-service sessions for staff ', only 35% admitted that they *did*. Some of the difficulties of achieving changes in attitude and improvements in practice emerge from the account which follows.

A coordinator

After leaving university in 1970 with an honours degree in electronic engineering, I worked in industry for two years. I then decided to pursue a career in teaching at the primary level, so I attended a one-year post-graduate course and obtained a Certificate in Education.

I began teaching at a small, inner-city church school in 1973, and it was here that my real learning began. Mathematics had always been of special interest to me but, as yet, played no more important a part in my teaching than the many other subjects a primary teacher normally takes.

The teaching at this school was 'semiformal', and after being given the usual 'forget everything they taught you at college' talk given to all new teachers fresh out of college, I settled down to teaching my children their reading, writing, and 'sums'. After a while, feeling there was more to mathematics than just 'sums', I applied to attend an in-service course of primary mathematics. My headmaster, who always encouraged his staff to attend training courses, released me one day per week in order that I might further my education in this field.

The course involved much practical work and gave me my first experience of 'modern mathematics'. At the end of the course I was very impressed by these new methods, and convinced that this was the way to teach children mathematics. I rushed back to my classroom, full of enthusiasm, to try out these new methods; but in my naïvety I had expected far too much, too quickly, and soon became disillusioned when things did not go as smoothly as I had expected: the children had not been previously trained in these methods, I had not fully grasped the concepts involved, and nobody else was in the least bit interested.

The problem I have always experienced when attending in-service training courses, especially in mathematics, is that whilst on the course I am surrounded by teachers who are interested in the subject, in finding out and experiencing and applying new methods; whereas, on returning to school, I find I am alone, fighting a tide of opposition born out of ignorance, insecurity, and single-mindedness. It is all too easy not to put into practice new methods and ideas when one feels such isolation.

During the last two years at this school I was asked to join a working party whose task it was to produce a detailed check-list showing the progression of skills and concepts in the area of number, from the infant level (age five–seven) to top junior level (ten–eleven). Throughout this time I learnt much through practical involvement in this work, and my interest and understanding of mathematics grew. It was at the end of this work that I felt qualified to apply for the position I now hold: head of mathematics in a large, three-form entry junior school.

At first I saw my role in very superficial terms; I thought that the mathematics teaching within the school would continue smoothly, without my interference, and that my job would be to

1 keep staff informed about relevant courses;
2 pass on new developments in mathematics teaching;
3 circulate articles, press cuttings, etc.

In fact, I expected a purely administrative role, which all seemed so easy until I realized, rather quickly, that:

1 Mathematics courses were the last thing they would want to attend – 'That's your job!'
2 'New methods? No, thank you! Haven't you heard? There's a big move back to formal methods as the government is very concerned about the drop in standards since these new methods came in.'
3 Articles were hastily read, signed, and forgotten.

No, it wasn't going to be easy. I needed to assess the situation and rethink my role. I had to make myself aware of the constraints such as the policy of the head, the attitude and experience of the staff, and attempt to devise a plan within these constraints.

During the first few months I looked at the effectiveness of the scheme of work, the monitoring system, resources, testing procedure and methods. The scheme of work I found quite satisfactory; up-to-date, but rarely used – teachers much prefer to rely on their own experience than on a document produced by someone else. The monitoring system I will go into later. As far as resources were concerned there was an adequate though unused supply of apparatus, together with a reasonably modern set of textbooks. The testing procedure was random and too subjective for my liking, and the methods were sound but out-of-date. The quality of teaching, however, was good and I realized that any changes I made concerning methods would have to be subtle if I was to succeed in updating such methods.

Every half-year all record cards were submitted to the heads of department. This gave me the opportunity to look into the monitoring system more closely. For each year-group I had to go through about 100 almost identical record cards. Each card showed a list of topics covered each month by each child, with comments by the side such as 'good', 'slow', etc. I could see no possible use for these cards in their present state. I consulted the teachers about a new monitoring system, as I felt they were wasting valuable time completing the old record cards. We decided that we needed a system which gave us useful, positive information about each child's progress. Ideally, knowing how useful my practical involvement on the working party had been to me, I should have liked to work with the staff on producing the new system. Unfortunately it was decided that, as head of mathematics, this was for me to do alone.

So, what would my objectives be? Could I use this as an opportunity to bring about changes in teaching methods, and to coordinate mathematics teaching throughout the school?

I am fairly convinced that when modern mathematics took hold during the 1960s, it did so before the majority of teachers had fully grasped the concepts involved; and without this understanding of the basic principles mathematics teaching became a haphazard activity. Mathematics is a subject to be taught progressively; that is to say, the order of teaching skills is vital and fairly rigid. So my aim in devising a new monitoring system was threefold:

1 to monitor progress in a meaningful way;
2 to reinstate content progression (always a strong feature of the traditional system);
3 to give staff a set of common objectives.

At the same time I needed to stress the idea that, as well as 'content' progression, there must also exist a 'method' progression; children require an initial period of concrete activity to develop the mathematical ideas at each important stage in the 'content' progression.

I decided to use a check-list method of recording progress, so I set about listing the progression of skills to be covered in each area of the junior mathematics curriculum. This was a mammoth task to attempt alone, but it was made easier by the fact that I had already been involved in producing a check-list for number. Rather than include every skill and concept to be mastered, I chose a series of 'stepping stones'. There would be a check-list for each child, which would be kept up-to-date by his teacher and passed on to the next teacher the following year. Progress would be made along several fronts at the same time, and the teacher would indicate whether a child had had contact with a skill or had understanding of that skill. In completing these check-lists the teachers would be regularly and repeatedly exposed to the progression of skills and concepts in all areas of the mathematics curriculum. I hoped that this would affect the order in which they taught the skills. I then had to convince staff that method progression was also vital; the use of structural apparatus does not stop when a child leaves the infant school.

I felt that an in-service training course of my own might have been the answer, but at the time I did not have the confidence, experience, or time to prepare such a course. Then, as now, my normal, everyday teaching work-load must take priority, thus limiting the amount of extra work that can be undertaken as head of mathematics.

I approached the problem by supplying my colleagues with a number of up-to-date booklets covering the whole spectrum of junior mathematics, which I obtained from the curriculum development leader in my local education authority. I supplemented these with a selection of sample work-sheets and my own handouts, hoping that this would stimulate staff discussion.

It was at this point that I set out a number of short-term objectives:

1   to ensure successful implementation of the check-list method of recording progress;
2   to implement a testing system to cross-reference with the check-list;
3   to encourage teachers to teach mathematical skills in a relevant progression;
4   to update and standardize methods.

We have now achieved smooth running of the monitoring system, ensuring that we have an accurate record of each child's progress in all areas of the mathematics curriculum. This is largely due to the hard work and excellent cooperation of the staff. I do feel, however, that we need to discuss the effectiveness of the system in relation to the time teachers spend on it. There is now a more ordered approach to the teaching of skills and concepts; the staff are better informed and the unification of methods is progressing slowly. Unfortunately the testing system proved to be impractical, but I am currently a member of a working party involved in this type of testing.

I am at present building up a resources centre which will contain booklets, magazines, and other literature full of ideas and suggestions on how to make mathematics a more interesting and enjoyable subject for children.

After two years as head of mathematics I am now very aware that it is no easy task influencing twelve individuals with differing attitudes, abilities, and experience. I am convinced that direction from me is not enough; teachers need to be closely involved in any decisions that will affect their work. Only through discussion and the exchange of ideas will we make progress and improvement.

Bearing this in mind, my immediate future aims will be:

1   to develop a more 'workable' scheme of work;
2   to build up resource material that will be used effectively;
3   to establish liaison with our secondary school.

Above all, I wish to create an atmosphere of cooperation in order that we might develop a more positive attitude towards mathematics teaching. [F]

*   *   *

Another approach is shown by the next writer. He is, however, unusual in having had a year in which he did not have responsibility for a class, and in which he could work alongside colleagues and with small groups of children. This gave him an uncommon opportunity to assess the state of mathematics in his school.

A coordinator

I am head of mathematics in a fairly newly opened primary school in the East of England. The school has children from the ages of five to eleven. We have 350 children on roll, and they come from families which are generally interested in the children's education and encourage them in very many ways. They all live in their own homes and have few family problems, and the school has no discipline problems.

The staff consists of six very able infant teachers and nine junior teachers who vary considerably in their experience and expectations. Two or three of the staff will admit that mathematics is not their 'thing' but the others show various amounts of interest from time to time; none except myself would claim to be really enthusiastic about the subject.

My relationship with the infant part of the school is slightly different from that with the junior part. The infant staff have excellent teaching skills and all they need from me is backing and a boost to their confidence. I try to help them to develop their own mathematics rather than their teaching skills. With the junior teachers, on the other hand, as well as trying to give some of them confidence, I am trying to help them with their teaching skill and technique.

Throughout the school there is a scheme based on *Mathematics for Schools* by H. Fletcher; there is also a lot of supportive material such as apparatus and books and workcards. I have been at the school for three years now, and last year we tried something very new. Instead of having a class of my own full-time and teaching all the subjects to them (as every other teacher in the school does) I was able to teach mathematics for 80% of my time to groups of children. This was possible because of a change in the staffing ratio of the school. There are two classes in each year-group, and I removed from each year-group a set of children deemed to be the least able and another set of the very best. I took each set two or three times a week. I also dealt with all the mathematics TV programmes, taking all the children, and following up the programmes with the whole year-group.

A group of thirteen nine-year-olds arrived in my room with problems one day. They had been struggling with addition involving tens as well as units and they just could not cope. This problem must be easy enough – not enough use of practical material, I thought. 'Do their teachers know about Dienes' apparatus, Cuisenaire, even an abacus?' I thought I had made all the apparatus quite clear to all the staff. I only had thirteen children, so I decided to *talk* with the children about adding up various articles. This gave me some idea of the problem – they could not talk about it. 'So let's have a great heap of Centicubes – surely they can count.' I gave some to each child and asked them to count them. They eventually all managed it, and I had a breakthrough in being able to talk to them all individually about their counting. 'Any ideas for arranging these Centicubes so that they look a bit neater?' I asked. Well, you would think I had asked for the moon.

'Let's put them in a straight line.'
'Let's put them in colours.'

This was all they could suggest. I thought I must be patient, so I said 'O.K. Sort them into colours.' They did so, some quite quickly, some very slowly. Most children also put them into straight lines. I had a quick look around.

Some of the children had very small numbers of counters in some of their lines – good. 'Kim has got three reds – could anyone make it up to ten?' Those of the children with small collections of reds brought them forward to join with Kim's so that they could count them to see if they made ten . . . and so it went on lesson after lesson.

We did get somewhere in the end. The children enjoyed coming. They became free to talk about their problems, and I had enough time and no pressures on me, so that I could listen and wait for them. A year later they still find odd moments to talk to me about their mathematics, maybe when they are working in the library or having dinner.

Now it is the start of a seventy-five-minute session with the ablest of the top year. I plan to talk to them about the meaning of the word 'operation'. I will talk about surgical operations 'changing the patient' and draw a few parallels. There was also that article in *Mathematics Teaching* recently about combining rows and columns together to give rectangles at 45°. The article talked about the operation of calculating the number of dots in any rectangle. The children were certainly very attentive, since they were fed up with ploughing through some of the books which form the main core of the curriculum. It is difficult to get your own ideas completely conveyed to all teachers, and there are some who do not see the potential of these particular children. The children are looking for something new, so here we go with the spotted patterns.

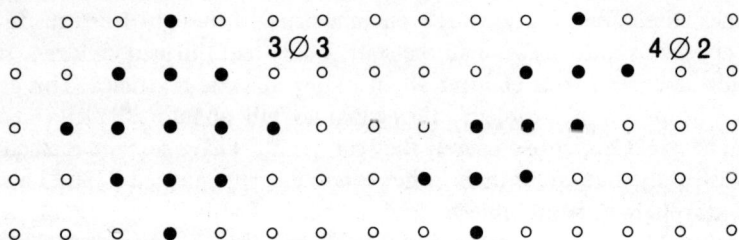

I hand out sheets of spotted paper (one of the most useful pieces of material in my cupboard) and say that we are going to draw rectangles round the dots, but not in the usual fashion, at 45° instead. 'How many dots inside?' I ask. There is a slight pause, and then

'It's a bit like multiplication but we get the wrong answers', says one lad.
'Are they wrong?' I ask.
'Well, no – because I've counted them . . . you just get different results.'

Ø	1	2	3	4	5	6	7	8	9	10
1	1	2	3	4	5	6	7	8	9	10
2	2	5	8	11	14	17	20	23	26	29
3	3	8	13	18	23	28	33	38	43	48
4	4	11	18	25	32	39	46	53	60	67
5	5	14	23	32	41	50	59	68	77	86
6	6	17	28	39	50	61	72	83	94	105
7	7	20	33	46	59	72	85	98	111	124
8	8	23	38	53	68	83	98	113	128	143
9	9	26	43	60	77	94	111	128	145	162
10	10	29	48	67	86	105	124	143	162	181

We talk about the new operation and one of the children decides to call it 'axitating', and the name sticks. Some of the children decide to draw all the rectangles which produce '2 axitating N' – a sort of 2 times table. One girl goes off on her own and does some fairly big rectangles, and decides to compare her axitating answer with ordinary multiplication. She calls me over and starts talking about these new rectangles having just the same number of dots as in an ordinary multiplication situation, but there are some extra shorter rows in between. I go from group to group, wishing I could record all that is happening. I suggest that we really could do with a sign for the operation; we decide upon Ø. A group soon produces an axitating table square. Someone else looks at 'square numbers', 7 Ø 7 for instance, which equals 85. The girl who was working alone comes up with the idea that you do not have to draw anything, as all the other children are busily doing:

'You just do the multiplication and repeat it with 1 less for both numbers: 12 Ø 8 = (12 × 8) + (11 × 7)= 173.'

A couple of boys excitedly tell me that axitating triangular numbers are the same as ordinary square numbers. A bell goes, but no one wants to go.

'Can we come in at dinner time?'
'Oh yes, alright.' I'm delighted really.

After nine sessions we had gone quite far. The school year was ending, and we were very pleased with the time spent. The children got a lot out of it that would be very difficult to put into words. Some of the staff thought we were mad doing 'space maths'. Pure, creative mathematics with children of eleven, and what a relief from classroom geometry, algebra, and endless decimals and fractions. At the end a child remarked, 'I wonder if they will ever invent a calculator to axitate?' I replied that it should not be too difficult.

Now it is one year later and I am back in the usual role of having a full-time class and trying to help all the teachers. The infant teachers say they want to know more about Logiblocks. They have had the apparatus for some time, but feel they are not getting the most out of it. This is not too difficult to arrange. We will have some practical work sessions during the last half-hour of the day once a week. The infant staff finish earlier than the juniors and the other teacher in my year-group will take all the children in the year for a story. There is new apparatus arriving regularly. I usually put it in the middle of the staff room table during break and involve some of the staff in playing with it. Then someone usually goes off to try it with the children.

The headmaster asks all the staff for forecasts of their lesson notes once a fortnight. In these, the staff put down details of what work they are planning to do and how they plan to go about it. He will often refer individuals to me or discuss some of their ideas with me. I then try to find an odd moment, perhaps during hymn practice, to talk it over with the teacher. We have a new teacher who has little experience as yet. She has said that the children seem to be a little shaky on the use of decimals. I'll be a few minutes late back to my class (the teacher next door is very good at keeping an eye on my class for me if I am late back); I find a few supplementary books and suggest using the Dienes' apparatus but using the 'unit' piece as one hundredth, the 'tens' piece as a tenth, and so on; her eyes light up as she gets the idea. However, I'll have to check that she is not going too quickly.

When the weather is bad my class goes each day along a corridor which is open to half the classes in the junior department. When I pass through with them, it is quite common for a teacher to talk with me about the work that a particular child is doing, or about some problem she is dealing with, or directly ask for help. It also gives me an opportunity to enjoy with my colleagues the many good things which they are doing with the children.

We have a very individual approach to the children. No two children do exactly the same pieces of work. The opportunity to work with small groups last year really enabled me to get to know how individual children 'clicked' mathematically. The effect this year has been to make the help which I try to give the other teachers much easier and more constructive. [G]

*   *   *

The next contribution is from a teacher who has a good knowledge of mathematics, but does not hold a post of responsibility for it.

A primary school teacher

The post involves much more than just ordering the mathematics equipment. There is need for continuous critical appraisal of the mathematics curriculum and procedures used in the school. Other members of staff have to be supported and advised on the best procedures for mathematics teaching – use of equipment, trying new approaches when others have failed, injecting new ideas, suggesting resources. . . . It requires confidence in one's own mathematical abilities and the ability to inspire confidence and enthusiasm in others.

Thus, the post needs experience and a certain amount of expertise not only in the mathematics taught in the primary school, but also in mathematical education, that is, in the way in which it is taught, what its aims are, etc. You have to take mathematical theory, such as that of Fletcher and Dienes, and present it to staff in such a way as to convince them of the value of these ideas. It must be *seen* to be of value to the children while being practical in teaching terms! It is this coalescing of the theory and practice of mathematics which is the most difficult thing to transmit to others – this is one of the main reasons why I think there is such a need for in-service courses. I think the difficulty of convincing people of the value of modern schemes is due not only to the nature of mathematics, but also to the fact that many have had such poor experiences of mathematics themselves. Why go about it in such a roundabout way? Why waste so much paper? Why introduce high-flying things like probability? We didn't do this when we were at school . . . and so on. Mathematics has such a stigma about it that individuals are quite naturally dubious about new concepts and new approaches. The job involves convincing people (as opposed to forcing them) that these concepts are of value, and that the methods used are the most efficient ones we know for learning them.

I see two main problems confronting the person with responsibility for mathematics. They are, first, general apathy (and sometimes animosity) towards mathematics, and secondly the insufficient knowledge that teachers have about mathematics and its teaching.

The natural and desirable questioning of new ideas and techniques presents problems to the teacher responsible – especially if he is a non-mathematician, which is highly probable in the primary school. Often these questions are both subtle and pertinent. The teacher with responsibility finds himself in the same boat as his colleagues. He has to understand these ideas, translate them, and explain. The explanations have to be convincing, but not condescending – perhaps a problem of management. Failure to explain might lead to apathy. (Why bother asking – he only knows as much as I do!) Not explaining in the right manner might lead to friction. The problem is compounded when he has to advise older or more senior members of staff, who are often more experienced. A similar problem occurs when he has to ensure that schemes, apparatus, etc., are being used. What are the tell-tale signs? How can the right use of apparatus and resources be determined without spying? Even if he knows that certain members of staff are not 'pulling their weight', what can he do about it?

These points emphasize (for me at least) the importance of having some sort of in-service course for teachers with such responsibilities. The task is an extremely difficult one. It requires the nonspecialist to act as an authority in an extremely volatile area. One of the big problems is enticing teachers onto courses, since there is such a stigma attached to any mathematics course that many are frightened off. It is for this reason that I would suggest that such courses should be very practical, although theory is definitely needed (in order to answer some of the awkward questions mentioned).

A course for headteachers to coincide with the above might also be valuable. It might give the teacher more backing, since he will have the weight of authority behind him. It might also help if he knows that there is someone apart from himself who at least has the same idea. The head is the best person to have. I am thinking of perhaps a three-day course. [H]

The same teacher went on to describe his ideas about in-service work in mathematics in general.

My view of in-service work is relatively simple. I think it should be directly related to classroom practice. Short courses can be valuable but they take the teacher out of the classroom. I think this is a good thing in that new ideas can be 'tried out' by the teacher without the pressure associated with working in school. On the other hand they leave the actual application of ideas in the classroom to the teacher. A fair comment would be that this should be so, it is after all the teacher's job – he is supposed to be a professional. The problem with this, as I see it, is that one assumes that the teacher has an expert knowledge of every subject, and that if these ideas do not work he has the time and the expertise to find a new way of tackling the problem.

The problems facing a teacher confronted with 'new approaches' to mathematics are of a type which necessitate some knowledge of what principles are involved. It's my experience that teachers are easily discouraged and feel insecure. 'I was never very good at it', 'I never did understand' are common remarks. Short courses might provide an initial stimulus but they then leave the teacher to his own devices. It's impossible to transmit a whole conceptual framework, a way of thinking about mathematics, in a few days. Interest rapidly dies when problems become apparently insurmountable. Such courses lead to the type of thinking reflected in the comment 'Every expert can convince you that his subject should have the most time.'

If teachers are sceptical of 'new methods' in general, small wonder that they are sceptical of new ways of teaching mathematics. After all it is a subject which is traditionally difficult and now, to everyone's bemusement, 'obscure' and 'difficult' areas such as statistics and probability are introduced. In short, 'new approaches' such as SMP or Fletcher demand of the teacher knowledge of the nature of mathematics as opposed to just the basics, the four rules. The problem then is one of convincing and providing enough mathematical background so that the teacher can cope with problems in the classroom. I firmly believe that if a teacher is interested and feels a part of the new process he will teach it well. A couple of years ago a number of staff I work with were greatly impressed to find that their children's scores on a particular type of mathematics test had improved. The scheme being followed – Fletcher Mathematics – up to that point hadn't really taken off. If there was a time that it could be said to have 'taken off' it was then. As a fourth-year teacher I can see the benefits of this now. Children approach mathematics with an 'I'll have a go' attitude. The points I'm really trying to make here are: when teachers become secure and confident in their ability to teach mathematics they teach it well and this process takes place over a period of time, and it occurs within the school situation. Quite honestly I doubt if this would have taken place had one teacher gone on a course and come back and explained what he had heard to the staff. The approach we followed was one involving the whole staff which allowed an interplay of ideas. The problems of implementation of methods were discussed in the school and some of the solutions and suggestions of how the problems might be overcome came from a visiting tutor from the local college, a sort of 'guest expert'. As the tutor became known the staff used her as support when difficulties arose. More important, perhaps, is that staff saw that these new ideas could be taught by the person who put them forward. I think this is important because it is all very well putting forward new ideas but nothing is more convincing than seeing them taught. Teachers respect other teachers who teach well and are more likely to listen to their ideas. It is this type of in-service work which I think is valuable; it provides support and is practical while conveying the principles of the subject. It helps build up confidence and security in the sense that someone is there who knows what they are doing and can provide support. [H]

CHAPTER 4

IN THE SECONDARY SCHOOL

A teacher's working time oscillates between alternating phases. In one, he (or she, of course) is usually closeted in a classroom as the only adult among twenty or thirty children; then, at 'natural breaks', he switches to his other role as a member of 'the staff', which may consist of anything from about twenty to a hundred or more colleagues.

It is a long-established practice in English secondary schools to subdivide the staff into departments according to their teaching subjects. The post of 'head of the mathematics department' is an official appointment, for which the holder gets extra pay – and takes on extra responsibilities in return. He is often appointed from outside the school, and will be selected for his personal qualities as well as for his experience and knowledge of mathematics. His job is to represent the subject within the school, to provide the organization for effective teaching of mathematics through the school, and to lead the team of mathematics teachers.

One problem in primary schools is that few of the teachers have a strong knowledge of, or interest in, mathematics; worse, many have no confidence in their ability to cope with the subject once it strays from the rigid routine of the textbook. At least secondary schools should not have that problem:

most members of the mathematics department have chosen to teach the subject, and at the very least might be expected to find it congenial. What more is needed, then, for the school to get the best from its mathematics teachers?

Three complementary answers to that question suggest themselves:

1   They should work out common goals for the mathematics teaching within the school.
2   They should keep their knowledge of the subject active and growing.
3   They should interrelate effectively as a team.

These are the themes which are worked out in the three accounts in this chapter.

An ex-head of department

'What is the most important piece of equipment a mathematics department should have?' I was asked.

'An electric kettle,' I replied without hesitation.

It needed a little explanation for my audience, who were teachers on a course for 'aspiring heads of mathematics departments in secondary schools', and I allowed myself the luxury of reminiscence, from my last school, before I became a full-time provider of in-service training.

We, my department and I, had been a gregarious lot. We sat together at coffee and at lunch and talked about teaching. We bored the pants off anyone else rash enough to sit with us. In the early days of 'modern mathematics' we met every two weeks in the evening at my house, and after soup and sandwiches we got down to the teaching of sets, vectors, matrices, or whatever topic was coming up next. If I gave a lecture at any local meeting, say, of the London branch of the Association of Teachers of Mathematics, my department would be sitting in the front row. If the meeting was held at our school then everyone would take part in some way.

In time the pressures of ignorance decreased, and also the younger members of the department developed those domestic responsibilities that make evening and weekend meetings more difficult to frequent, but the enthusiasm remained, and so did the gregarious instinct. The best time to satisfy it now seemed to be that magic hour at the end of school when the children have left, and teachers have that strange unwillingness to go home before they have relaxed and unwound.

We became naturally drawn to the main mathematics room, which was on the way down from the other rooms and was where we kept the central supply of equipment. We gradually arrived. We talked. We shared our day's failures and successes. We picked up ideas on mathematics or on teaching and developed them or argued about them. We raised our problems about tomorrow's lessons and helped each other to solve them. We prepared both ideas and materials. An hour or an hour and a half later we left. The nearby streets were clear of children and we were clear of some of our anxieties.

In our case the electric kettle was necessary in order to satisfy our physical thirst, at the same time as our informal meetings satisfied our other needs. I suppose I fancied that this process might also work the other way round – that the kettle would provide the coffee or tea which in turn would provide the attraction to meet together at appropriate times, from which would develop the gregarious spirit and the interplay of ideas that continually revivify the work of any department.

It also meant for us a harmony and an ability to work together that carried into the classroom. Perhaps our own source of liquid refreshment was more convenient than the central one in the school staff room, but in their free periods our teachers, rather than going there to mark books, would go into someone else's classroom to watch, to help, to join in, to take over. (This was how team-teaching began – as team-learning.)

There were more celebratory occasions when, perhaps at a different venue, the drinks were stronger. At one time the advertisements I prepared for new teachers always ended with the sentence 'Ability to drink beer an advantage', but it never got through the censor!

The department that drinks together, thinks together. [I]

*   *   *

A head of department

I have a department of ten mathematics teachers, three having Open University degrees, one studying for a degree, and five college-trained. Until 1976 most of the work of the department was for the CSE examination. In 1976 the school began to take pupils across the whole ability-range and it became necessary to consider O- and A-Level courses in the school's chosen syllabus, that of the School Mathematics Project. I joined the school in 1978, having previously taught in a school with established courses in SMP A-Level mathematics and further mathematics.

We have a meeting of the department once a month in which we discuss the problems that are arising with the work, exactly what the content of our curriculum should be, and in what direction we should go in the future. It became clear from these meetings that one of the greatest difficulties lay in knowing exactly where the emphasis should be in each topic and in the course as a whole. Associated with this were the decisions about how long to spend on each topic, and how important it is that the topic is understood by the class – whether it is a building block or just decoration as far as future work is concerned. It was suggested at one of the meetings that I might find some way in which I could make my experience available to all the department.

My first thought was to try to put this experience down on paper – but I soon abandoned this as a method. At the next monthly meeting I suggested that I should teach them all the A-Level course, just as if they had recently completed their O-Levels. The response was enthusiastic and I began the first of a series of weekly meetings with all the department present except the one member with A-Level teaching experience. We shall be continuing these meetings until we complete the A-Level course and then review the situation.

I usually take each topic from its origins in O-Level and proceed as soon as possible through the A-Level work, indicating how I tackle each chapter with a class, which parts of the chapter are most important, and where I have experienced difficulties. I usually finish each session with some class exercises which are completed at home. Some of my colleagues are using these 'lessons' to prepare for teaching A-Level themselves, while others take items for use with O-Level additional mathematics classes or merely to see what impact these can have on their teaching in the eleven–thirteen-year-old age-range. All of them feel that it will help them to see where the O-Level course is leading, and to adjust the emphasis of their teaching in the lower classes so that they can best help the pupils who are potential sixth-form students.

From my colleagues' point of view, one of the greatest difficulties is finding an hour after school every week, as many of them have other educational responsibilities such as pastoral care and out-of-school activities which can cause them to miss a lesson. They are all too well aware of the difficulties caused by absence from lessons! As a department we are also involved in other projects – for example, meetings about CSE course-work, liaison with industry, liaison with primary schools (where we are trying to reach a common understanding about the mathematics for the nine–twelve-year-old age-range), internal examinations, staff meetings, and marking homework. It is certainly a dedicated teacher who can still find time to spend an hour a week being taught mathematics.

We now have three O-Level groups in each of the third and fourth years, and the staff who are teaching these groups say that they already have an increasing awareness and insight into their work and that it will benefit potential sixth-formers. For example, they now realize the importance of work on vectors and functions at this stage, having seen how large a part they play in the A-Level work.

From my point of view, the greatest difficulty is to find a suitable level at which to aim my 'lessons', since the experience of the group ranges from CSE to degree level. This is the reason why I decided to start from inside the O-Level work before going on to the A-Level books. It is also quite clear that those staff not regularly teaching an O-Level group or above are finding the work difficult. I admire their perseverance and I am grateful for an enthusiastic team who take such an interest in their job beyond the walls of the classroom. [J]

*   *   *

A head of department

My own personal view of in-service training and the role of the head of the mathematics department can be analysed by emphasizing four different aspects of the problem. I have named each emphasis after a famous mathematician so that a light-hearted comparison can be made between it and a characteristic piece of mathematics of its adopted mathematician.

I have associated the first aspect of the role played by the head of the mathematics department in in-service training with the name of Eratosthenes. In this approach the head of department acts as a sieve and periodically sifts through current research and practice in mathematical education. The criterion on which a synthesis is made is summed up by asking two questions: 'Is a particular piece of research relevant to the needs of my pupils?', 'How are the results of the research going to influence future teaching policy within the department?'

However, this sifting process which characterizes the Eratosthenes approach is not the exclusive domain of the head of department. All members of the department should be given the opportunity to play an active part. One way of doing this is to encourage members of staff to read and comment on relevant journal articles. Three journals which I have found particularly useful in this respect are *Mathematics in School*, *Mathematics Teaching*, and *Educational Studies in Mathematics*.

A departmental seminar under the leadership and/or chairmanship of the head of department is one way of putting this approach into practice. It has the advantage that it gives individual members of staff the opportunity to play an active part in discussions which may lead to a change in departmental philosophy. It certainly tells the head of department what his staff consider an appropriate gauge size for the sieve to be used in a particular synthesis of current research.

The second aspect of the role played by the head of department in in-service training I have associated with Euclid. It emphasizes the logical approach a head of department should adopt when he considers a request made by a member of his department to attend an in-service mathematics course: 'How will the member of staff benefit from the course?', 'What will be the professional gain to the department?' The head of department also has the option of recommending particular in-service courses to individual members of staff. Thus, this second approach emphasizes in-service courses organized by external bodies.

An attempt to answer the question 'How will the member of staff benefit from the course?' must be concerned with the way an in-service course influences the personal philosophy of teaching mathematics held by an individual teacher. Is his classroom teaching going to improve as a direct result? Will he come away at the end of the course realizing that his traditional way of teaching long multiplication to 4Z may be inferior to one based on Napier's bones? Indeed, should he be spending so much time teaching long multiplication to 4Z anyway? Will he be convinced that in addition to encouraging his pupils to give the automatic response of $\sin x + c$ when presented with the stimulus $\int \cos x \, dx$ he should also try to develop in them a conceptual understanding of the problem.

A direct answer to the second question 'What will be the professional gain to the department?' is not possible when only short-term gain is under consideration. Any general benefit will be long term and will take some time to make itself apparent. Is the existing departmental philosophy of the learning and teaching of mathematics going to be enhanced? Will the member of staff return sufficiently inspired and impressed by the course to give support to his head of department's belief that an element of continuous assessment is vital in any structured scheme of teaching mathematics? Or will he come back convinced that all classes should be mixed ability irrespective of the peculiar difficulties that may exist in different schools?

The third aspect of my analysis I have associated with Einstein. It is in this approach that a head of department has to evaluate the professional expertise of his teachers, especially probationary teachers, relative to a set of departmental guidelines. I have found it useful to classify such guidelines into five sections. Each section, which in practice overlaps with each of the other sections, has a check-list of three questions, which demand a yes/no response, and I would like to suggest that the relative professional competence of a mathematics teacher is in direct proportion to the number of *yes* responses. Naturally, an allowance for the weighting appropriate for individual questions has to be taken into account in any numerical evaluation. The fifteen questions, subdivided into five sections, are as follows:

1 *The teaching of mathematical concepts.* Does the teacher begin his development with concrete examples and allow sufficient practice to be gained before proceeding on to mathematical abstraction? Is the material being introduced to the pupils at an appropriate conceptual level? Is the teaching sequence organized into simple hierarchical steps?

2 *Lesson preparation and construction.* At the beginning of each lesson does the teacher make sure that his pupils do have the assumed prerequisite knowledge? Does the main part of the lesson follow the example-rule-example pattern? At the end of the lesson does the teacher bring the class together as a unit to give a summary of the lesson and to make a brief analysis of the behaviour modifications which have taken place in his pupils?

3 *Chalk and talk.* When the teacher is addressing the class is his intonation such that all the class are listening? Is the development of his argument summarized legibly on the blackboard? Does the teacher clarify any of his explanations by the use of visual aids?

4 *Information and its communication.* Does the teacher try to get information from his pupils by asking a large number of questions during his teaching? Does he attempt to grade his questions so that all his pupils from the weakest to the brightest have at least the opportunity of answering some questions? Does he make any attempt to increase the understanding of individual pupils by providing extra help during the problem session part of the lesson?

5 *Relationships within the class.* Is the general relationship between the teacher and his class such as at least to maintain if not improve pupil motivation to learn mathematics? When pupils are engaged on mathematical problem-solving is the general noise level acceptable to other members of the class? Does the teacher enjoy the respect of his

class to the extent that class control does not feature as a major issue?

The fourth and final aspect of the analysis I have made of the role played by the head of department in in-service training I have associated with Euler. In this approach the head of department tries to find a relationship which unites each of the three separate approaches already discussed. One way of arriving at such a unifying relationship is to consider one general facet of the role of a head of a mathematics department. Any head of department must try to develop a common departmental philosophy for the learning and teaching of mathematics. A department in which each member agrees with its main aims and objectives and strives to keep inconsistencies of methodology to a minimum will be successful in its teaching. A department in which some members of staff disagree with its general philosophy will be less successful because it is upon such human characteristics as the attitude and responsiveness of individual teachers that good classroom practice ultimately depends. It is my hypothesis that in-service training is the unifying relationship that a head of department and his staff need to make use of when they attempt to develop together a common departmental philosophy for the teaching and learning of mathematics. [K]

CHAPTER 5

LINKS BETWEEN SCHOOLS

'All change'

We have a problem. Once, twice, or even three times during his school career a child takes part in a mass migration – from primary school to secondary, from middle school to upper, from high school to sixth-form college, or whatever (see the diagram on page 153).

In some LEAs the process is tidily organized, with all the children from a small number of 'younger' schools going to one 'older' school. Elsewhere, especially where a wide choice of schools is offered to parents, there may be no recognizable pattern of redistribution.

But either way, we have a problem; having more schools involved just makes it bigger.

If all schools followed the same mathematics course and used the same textbooks, if all children developed at the same rate and never missed school, if all teachers shared a common view of mathematics . . . then perhaps the problem would go away. (Perhaps it wouldn't.) But living as we do in a world peopled with real teachers and real children, the best we can do is to turn the problem to advantage. And few things can contribute more to a teacher's professional development than the need to work out solutions to professional problems by common consent. Justifying our own practices and attitudes, contrasting them with those of colleagues, examining in detail alternative materials and courses, and all in relation to children for whom we have a responsibility – if we do not emerge from this as better teachers, maybe we are in the wrong profession.

How does it work out in practice? This chapter begins with descriptions of two models, both as it happens concerning 'transfer at eleven'. In the first the initiative came from a mathematics teacher, and the subject formed the basis of liaison. The second shows mathematics as just one component within a school-wide initiative.

A head of department

'When are you coming to test our pupils?' This telephone message from a primary school headmaster was the beginning of our liaison with primary schools. We had just opened that year as a new eleven–fourteen high school* receiving only a single first-year intake. What on earth was I going to do? I really had no intention of testing children coming up to me. At the beginning of the year the school had received an enormous bundle of information on each of the present pupils – general remarks and assessments of competence in basic subjects. It had been written and received in good faith, and I felt particularly sorry for the teachers who had compiled so careful a report; but it just wasn't very useful, and it just wasn't used – any of it!

The principal of my school thought that a test was a good idea in the current climate of opinion and suggested that I should at least think about it. After all, a neighbouring high school gave a battery of tests in mathematics; couldn't we have just one? We eventually decided on the form reproduced on pages 54–55.

The test was pulled from some research work by Margaret Brown at Chelsea College, and centred around the way in which children are able to *talk* about mathematical concepts and processes and their *understanding* of basic numerical computations. As a 'test' it was rather different from the norm, and I arranged to visit the primary schools to talk to the staff about its aims, and to explain why it might be useful. They were extremely interested and asked for a meeting afterwards to discuss the 'results'. I readily fixed one up and looked forward to the start of a really effective liaison programme.

The meeting was a disaster. Instead of an opportunity to discuss the broad issues and to tackle some fascinating individual problems, the schools wanted to know how they had done and whether they were better or worse than the rest. All their professional fears came to the surface and all the barriers came between us.

One school, whose catchment area was poor and who expected their pupils to do badly, was really pleased at finding they had done rather well. Another, which had a reputation for a good grounding in basic skills, was absolutely choked! Towards the end, I remember someone limply suggesting that we meet up again – but I knew that there would never be the same enthusiasm.

We have continued with the assessment 'test' and found it a joy to look at and read. For all its limitations, it is so revealing of the way in which children think. At the very least it serves as an illustration of what is important in mathematics teaching. Links with one school have developed along these very lines, and as a high school we have been able to help in a small way with resources. We have also visited them and they us, to see the ways in which children are learning.

In that first year I was naïve enough to ask all the primary school headmasters if I could visit their schools to see mathematics in action. I remember catching one head just as he was leaving school, and suggested that it would be nice to come and work with some children – could I come along and see mathematics at the school? 'No,' came his emphatic reply, 'you cannot!' To his great credit he went on to explain how awful mathematics teaching was at the school, and ended with the statement '. . . but I'm working on it!'

In general we have had little success with the more formal systems of liaison. Reports and assessment forms are not an effective means of communication in mathematics, nor are meetings between schools with their inherent formality and threatening nature. It is at the informal level that liaison can be effective. A recent local Saturday conference run by the Association of Teachers of Mathematics was most successful in bringing primary teachers, secondary teachers, and their pupils together and immersing them in mathematical activities. There is room for similar ventures at a more local level. Effective liaison involves teachers gaining a wider perspective of mathematics – a perspective through which a child's mathematical development is never restricted. Our liaison is a failure in these terms. However, there have been small successes along the way, and we at least know the direction in which to turn. [L]

*   *   *

Date	MATHEMATICS		Name	
School		Teacher		

1.	A bar of chocolate can be broken into 12 squares. There are 3 squares in a row. How do you work out how many rows there are?	12 + 3	3 × 4
		12 × 3	3 − 12
		6 + 6	12 ÷ 3
		12 − 3	3 ÷ 12

2.	A shop makes sandwiches. You can choose from 3 sorts of bread and 6 sorts of filling. How do you work out how many different sandwiches you could choose?	3 + 6	9 + 9
		6 ÷ 3	3 ÷ 6
		3 × 6	3 + 3
		6 − 3	3 × 3

3.	John has 28 records at home. Susan has 60 records. How do you work out how many more Susan has than John?	28 × 60	28 − 60
		28 + 60	60 × 28
		60 − 28	28 ÷ 60
		60 ÷ 28	60 + 28

4.	A gardener has 391 daffodils. They are to be planted in 23 flowerbeds. Each flowerbed is to have the same number of daffodils. How do you work out how many daffodils will be planted in each flowerbed?	391 − 23	23 + 391
		23 × 391	391 ÷ 23
		391 + 23	391 × 23
		23 ÷ 391	23 − 391

5.	John buys 7 packets of sweets. Each packet contains 17 sweets. How do you work out how many sweets John has altogether?	17 − 7	7 × 17
		7 + 17	7 ÷ 17
		17 × 7	7 ÷ 17
		7 − 17	17 + 7

An assessment test

1. Write a story for 84 − 28

2. Write a story for 9 ÷ 3

3. Write a story for 84 × 28

4. Write a story for 9 × 3

5. Write a story for 84 ÷ 28

A head of first-year in a secondary school

The recent 'Green Paper' on education issued by the government states that 'difficulties of transition within an area often arise because there is insufficient contact between teachers of secondary schools and those in their contributory primary or middle schools'.

I have been asked to write about this because others have judged that, in some way, our high school, with the cooperation of its contributory schools, is seeking to deepen that contact and to produce a constructive and continuing dialogue between teachers. Before writing of the progress we have made, it is only right to describe the background of the school, and to outline a little of my personal philosophy about the job I am doing.

We were fortunate to open a new, purpose-built high school in April 1975, an addition to the county's schools and not a replacement. It was built for 720 pupils (eleven–fourteen years old, eight-form entry), but initially we had only one year-group of some 200 pupils. The buildings were in one phase only, so both space and staffing were favourable during that term and, to a lesser extent, through all the next year. In many ways my appointment as head of the first-year, with responsibility for links with primary schools, was a golden opportunity – a new school with no 'history' to overcome, a catchment area of only three villages and just four 'feeder' primary schools in these villages.

Nevertheless, we have had to work hard together to establish the sort of liaison we believe to be most helpful. Links do not grow naturally, even in a favourable environment, and our contacts are the result of painstaking and diplomatic moves from both sides. There is a danger that the high school will be looked on as the initiator, but my experience proves that both need to have a positive approach to liaison. Ideas, and their adaptation and implementation, have come from both sides. The 'coordinating' role has often been left to me, but this account is by no means a summary of my efforts. Much of the credit (if that is the right word) goes to colleagues in the top junior classes, whose warmth and cooperation at every stage have been a constant source of inspiration to me and to the children.

In any child-centred view of education we need to pay regard to the previous experiences of the child as well as anticipating what lies ahead. In our LEA's system with three 'tiers' of schools there is a special need for close liaison. If the high school years are to be used to the full, they must build on the primary stage, and direct the pupils towards the upper school.

This concept of the 'family of schools' is increasingly talked about, and must weigh heavily on the minds of planners who seek to rationalize catchment areas. Continuity is a virtue we have often underestimated.

It is therefore one of my aims to seek to show children, and through them their parents, that transfer at eleven is not a giant leap, but rather a step in the continuing process of their education. That continuity is seen in the curriculum offered at both levels, but also in the philosophy and approach to education, the style of work, the tone of the school – the emphasis on values, respect for the individual, and so on.

One means of helping to bridge this gulf is through the understanding of the teaching staff. So we try to visit each other's schools at some length, and if possible to teach in the different situation down the road. The confidence of pupils and parents will increase greatly when staff can appreciate and explain the aims of the other school. In many ways primary staff are our ambassadors in the community, so I value their views and judgements very highly.

Education in no way begins at eleven, even though some secondary specialists may still hold that view. Nor does education stop at eleven, as some primary teachers would have us believe! Until we are prepared to meet and see each other at work, I fear those views will be perpetuated. If the child is seen as the centre of all we seek to do, then I believe, with goodwill and understanding, we will be able to improve the standard of his education. This has been my aim in undertaking the work of liaison outlined here.

When I first moved to this school, I had the opportunity to discuss my job in depth with the head, and his sympathy with my general aims has been invaluable. As a result, the school made a commitment to the work of liaison, and four of my ten nonteaching periods in a forty-period week are blocked together on one afternoon.

Between October 1 and April 30 each year, I try to spend these afternoons visiting primary schools. This is a substantial commitment by any standards, and it means I can spend one afternoon among every ten children due to move up the following September (three weeks with a class of thirty, in practice).

This allows me to develop the personal aspect of my work, which I believe must come first. Some written information will be passed on, to stop up the loopholes of a verbal system, but whenever possible I seek to provide a personal touch, whether it is in my contact with staff or pupils. No circular or brochure will ever fill the gaps in the way a conversation can, and the work I do at this level is perhaps the most important of all.

With this in mind, my specific objective on those afternoons is threefold. First, I seek to build good relationships with the staff of the primary schools. I always aim to spend half an hour of my lunchtime in the staff room, and often we discuss informally the progress of children already at the high school. At times there are topical educational matters we air together, but more often I simply share in the recreational chat of a primary staff room!

Second, I hope to talk to individual children about their particular involvement with the high school through elder brothers or sisters, older friends, or through their visits to the school for evening functions, etc. Often they have questions, and I always make time to give reasoned answers, allaying what fears I can!

Lastly, I try to examine the work of every child. In particular I hear everyone read, I see their current mathematics book, and I look at a project or creative writing book. I make brief notes while I am busy, but at some stage I always check these with the teacher, noting his assessment of the children's potential, as well as their particular strengths and weaknesses.

I have found these visits to be the central part of my work. Through them I am able to learn the names of many children before they reach the high school, and to discover quite a lot about them for myself. On top of that, I have been able to experience all our primary classrooms at first hand. I have 'felt' the atmosphere the children are used to, and I have grown to have great confidence in the primary staff. My full acceptance of them as professional colleagues, and their acceptance of me, is crucial. I feel able to accept their assessments readily, and I value their judgements.

Sometimes I am able to exchange with the fourth-year primary school staff, and on those occasions, when I have twenty or thirty children on my own, the afternoon is bound to be a little different. Usually I organize a special block of work for the afternoon, and I hope they will be able to get on with it alone for a lot of the time while I speak to individual children.

When this happens, I also arrange a programme for my primary colleague while he is at the high school. Occasionally this has involved him in teaching a small group of children, but more often he spends the time observing the children in various departments. Those colleagues who have exchanged with me have found it a particularly rewarding experience, and it does emphasize the two-way nature of liaison.

Apart from these regular half-day visits, we have arranged many other functions for staff, pupils, or parents. On the curriculum side, most of our heads of departments have visited the primary schools and gauged for themselves the approach to primary education in these schools as well as stating their own particular philosophies. Within the area of mathematics we have held meetings at each school to discuss the content of the syllabus in the two years surrounding transfer, and this dialogue is continuing. I am at present involved in discussions about one primary mathematics syllabus. My role is, of course, strictly advisory, but I am pleased to take it on. There have also been discussions regarding language development, and detailed sessions about our slow learners.

On an informal level, we frequently invite primary children to dramatic events; we link occasionally for concerts, and exhibitions or displays are often an excuse for small-group visits. Many primary school pupils first came into the high school to a film show or a country-dancing evening. Our resources and reprographic area is popular with primary colleagues, and we share the equipment we have.

At the primary schools I visit open nights, attend occasional concerts, and have at times participated in the school assembly programme. In all these exchanges we take the view that, time and pressure permitting, we want to be involved with each other's schools as much as possible.

As the end of the primary school fourth-year approaches, we organize some special events to ease the actual trauma of transfer. First, I arrange for each school to send us their fourth-year pupils, with two members of staff for half, three-quarters, or a whole day, as they wish. I arrange an interesting and varied programme to include some of the following items:

Lesson in gymnasium	More formal mathematics lesson
Cafeteria lunch	More formal English lesson
Tour of the school	Slides of 'design' department
Concert and sing-along	Slides of science work
Poetry reading	Visit to design centre in more depth
Lesson in drama studio	Questions

I hasten to add that I have never included all these in any one visit.

Second, we invite parents to an open evening where they listen to brief talks about the school as outlined in the school brochure, hear short musical and literary performances, and tour the school while children work in the practical areas.

Third, I ask primary colleagues to give me a written summary of each child, either on primary school record cards, or on a form I have devised, and in particular to advise me on certain things. I ask for small friendship groups that are known to work well, and information regarding bad relationships; I seek advice on the children's attainment in mathematics, and

set our first-year mathematics lessons purely on these guided recommendations; I need help concerning the degree of remedial cover required by
underachievers; and I ask for notes on medical or family circumstances as
well as the attainment of children learning musical instruments. Much of
this information I have discovered already, but I like to have this written
record in case of any omissions during my visits.

Once the children are established in the school, the first-year pupils are
screened by the basic studies department using a comprehension test, a
spelling test, and a piece of free writing done in half an hour.

At that stage, I make contact with the basic studies department, and there
is a remarkable correlation between our findings. Of the twenty-three
children I recommended for withdrawal this year, seventeen are given full
withdrawal in English, the other six all receive partial withdrawal, and there
is one other child withdrawn that I did not recommend.

I think that summarizes the complete range of the contacts we have had
with our primary schools in the first $2\frac{1}{2}$ years. The object of the exercise can
never be stated exactly, but I believe we are trying to develop a constructive
link between the schools in our area. The results of these links are hard to
isolate, and no doubt the response of our pupils is due to many contributory
factors. Nevertheless I believe that this work of liaison is partly responsible
for the positive and enthusiastic response of our pupils, and I feel sure that
my list of problems with first-year pupils is reduced greatly by the work that
has gone on in the previous twelve months. [M]

Can a catalyst help?

Linking schools is obviously a good thing; but initiatives can easily lead to
misunderstanding or even resentment. An 'older' school is suspected of
domineering; a 'younger' school is accused of trying to steal a march on its
neighbours. Just one individual with an unhelpful attitude can make cooperation impossible. Or fears that this might happen prevent the first initiative from being taken.

One possibility is to make available an individual from outside the
schools, to act as a kind of catalyst or facilitator. In the first of the examples
described, transfer between schools provided the impetus which soon
developed into interschool in-service support along broader lines. The
second began with broader and less specific aims, and some cooperation on
transfer was part of the spin-off in one area. In both cases, the 'outsiders'
were provided by local colleges, and they shared their time between work

with the separate schools and assisting interschool collaboration.

A college of education lecturer

Seven schools on the outskirts of a large town were linked through administrative reorganization:

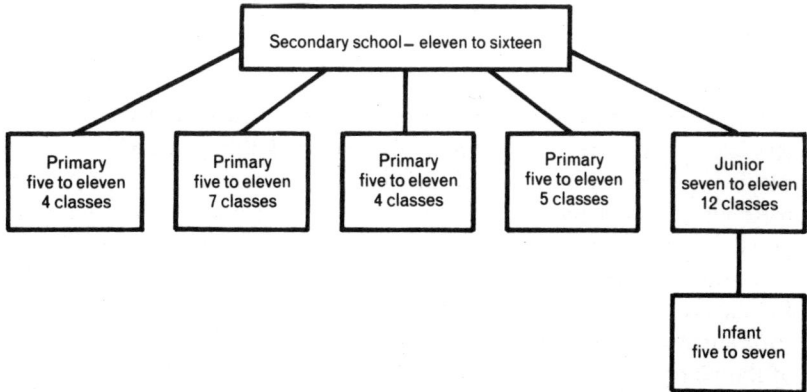

Formerly the pupils had transferred at eleven years to numerous schools throughout the county and town, but now the majority of these children will go to the same school. The first narrow issue of mathematical concern was raised by the secondary school: basic arithmetic skills and their layout. When new pupils had come from ten or even more primary schools, they had been taught a variety of different methods; now, with the main catchment area defined, the secondary school hoped that these discrepancies could be avoided and that all the children would be taught to record their calculations in the same format. However, it soon became apparent that the mathematical discussion would need to range more widely.

The primary school headteachers had long had the custom of meeting regularly. One of the primary schools had recently rethought its mathematics curriculum, and so the college tutor who had helped their school was now invited to work with all the schools and to facilitate their exploration of the wider issues emerging. The result was a three- to four-year in-service programme for the staff of all the schools.

Before describing the in-service programme, it is worth noting two aspects of the role of the 'outsider':

1   The facilitating character of the involvement. Most of the activities

could have taken place anyway, but the presence of the outsider provides a focus which is independent (and so hopefully impartial) and stimulating. It provokes activity and concentrates criticism.

2   The importance of acceptance of the facilitator by all the schools. It is essential, therefore, for this person to develop relationships within each school, being sensitive to the means through which this can be done and being prepared to give the time necessary for their development.

It has been noted that one of the primary schools had recently reviewed its curriculum; the other primary schools undertook the same task, so that all the 'feeder' schools were sure of their mathematical aims and their planned methods of achieving them. This stage of the exercise took at least two years, and actively involved all the teachers. Quite early on, a decision was made that there would not be a standard curriculum plan for all the schools. Practically this would have been difficult, since the schools had different equipment and books (some very recently acquired). The essential element of the plan (as envisaged by the facilitator) was to ensure that each set of teachers knew its own school's curriculum structure and was secure with it.

The approach to the task differed from school to school. The school which had already reviewed its curriculum had achieved this through a series of after-school meetings attended by the facilitator. Now they were actively implementing their decisions, and the facilitator was available for weekly consultations and classroom support. A display prepared by the teachers for a parents' meeting eventually helped teachers as well as parents to see the progression planned in the new scheme. Another school had embarked on an individualized workcard programme for the seven–eleven age-group. Some staff discussions were necessary, both to consider the progression within the scheme and to identify its links with the course followed by the younger children. The facilitator led these discussions and also helped teachers new to the school to develop familiarity with the programme within their classrooms. Lunchtime meetings and half-day closures of the school were used in the other schools to provide the opportunity for teachers to consider their existing mathematics curriculum. The immediate result was a greater awareness of progression (or lack of it) between classes, and a reconsideration of equipment needs and methods of recording children's development. The facilitator also met teachers in their classrooms, so providing an opportunity for individual discussion and for gaining a 'feel' of the individuality of each school. Secondary school contacts were not so easy to establish; but since it was essential to have a 'feel' of this school too, the facilitator taught some of the younger classes for a while when there was a staff shortage.

The next step was a series of interschool meetings:

1 Between the teachers of the top junior/primary classes, to compare the range of concepts and skills they listed as being attained by the 'average' children in their classes. There was considerable surprise expressed about the amount of agreement in this area, even though the resources used in each school differed. A common core was speedily agreed.

2 Between teachers of the six–seven-year-old classes. It did not seem good or practicable to determine a common core for each age-level, since the pattern of work in each school was so different, but this age-group seemed a suitable transitional level to compare. The main value of the meeting, however, was the sharing and exchange between teachers, some of whom said that they had never had the experience of meeting teachers from another school engaged in similar work, in all their teaching careers.

3 Between top junior class teachers and the secondary school mathematics department. The main purpose of this meeting was to discuss the common core and its transition/transfer implications.

The project is not completed. Many areas still need to be considered. The secondary school is now rethinking its curriculum, and is trying to build from the common core; methods of preparing a mathematical profile of each child are being explored; no doubt, arithmetic recording formats will be considered – but the context has broadened. The facilitator has now left the area; will the schools still meet so that the mathematics curriculum continues to live and grow? It would be sad if the development stopped and ossified at this stage. Many teachers have experienced a mathematical maturation, but has there been sufficient impetus for them to continue to facilitate their own development? [N]

*   *   *

An adviser

In England and Wales teachers are trained in basically two types of institution: universities and colleges of education. The universities have a long tradition of academic independence which is fiercely protected, although much of their income is from government funds; the colleges of education and polytechnics are also almost totally dependent on government money, but the Department of Education and Science has tended to be more directive in this area. The distinction between the two sides of the 'binary' line has lessened with the introduction of degree courses for all teachers, but it is much easier for the government to control the colleges of education.

For this and other reasons, when it became apparent in the mid-1970s that the fall in the birth-rate implied a consequent cut in teacher training, the brunt of the cutback was in the colleges of education. Many were closed, many were reduced in size, many were merged with polytechnics, but all of them were required to devote about 20% of their teaching to in-service work which had previously formed a small, or even nonexistent, part of a typical lecturer's timetable.

In-service education can mean many different things. Most lecturers would prefer to teach high-level award-bearing courses leading to a B.Ed. degree or a diploma of some sort. If possible they would prefer to teach the course in the college rather than at a teachers' centre or similar outpost. While not wishing to denigrate such courses, I had a strong feeling that lecturers could also prove most valuable in helping teachers in the classroom when the schools were in session.

Thus, at the beginning of the academic session 1977–1978, I, as the county mathematics adviser, had persuaded three lecturers drawn from two colleges to devote approximately half a day per week for a year to working in schools in the county. My brief had been vague. I suggested that college lecturers had a poor image with teachers; an essential need would be to gain confidence and respect. The best way, it seemed, was to offer to do whatever the schools required, or requested.

The three lecturers each were given a fairly well-defined group of schools covering the age-range five–eighteen, and initial meetings were arranged to explain this novel service to the teachers. Of course, there was much suspicion on the part of the teachers and hesitation on the part of the lecturers.

All kinds of involvement followed. Some schools wished to overhaul their schemes of work; others wanted to discuss new textbook series. There were requests for demonstrations of apparatus and specimen lessons on new topics. Sometimes the lecturer worked alongside a teacher, sometimes he took the class, and at other times, usually after schools had closed, he ran short courses on requested topics.

In retrospect, each side was, initially, nervous of the other. Not all the lecturers were familiar with primary schools; the teachers were reminded of the tutor-student relationship when they were training. As the scheme developed, however, both groups gained much, leading to genuine partnership. It was possible, fortunately, to repeat the experiment in 1978–1979 in three different areas, but with the same three lecturers. Some tentative conclusions can be drawn:

1   The college lecturers themselves, as would be expected, gained considerably from the experience. The first-hand contact with the classroom has enriched their college teaching and improved relations between colleges and schools. Initially they did not know what was required, but they soon grasped the important (and delicate) nature of this 'low-level' work.

2   Teachers who would never attend a course outside school hours, especially at a teachers' centre, could not escape easily from the effect of the general discussions taking place in their own school.

3   The lecturers were neutral. They had not been commissioned by the local education authority. They did not work for the teachers' employer and thus could not, in any sense, affect promotion or other professional matters. In fact, the only condition attached to the provision of a lecturer's service was that it was used! (In one case it was necessary to move a lecturer to a second area when schools did not take up the offer.) Schools could thus admit to a lecturer that their mathematics scheme was deficient, or that some teachers were very weak, without the fear of this being noted officially. This confidentiality was maintained throughout.

4   The schools had access to more than a lecturer. They could borrow equipment and books and make use of the lecturer's colleagues on particular specialist points.

5   Some published material emerged, e.g., a set of objectives for schools to work towards when transferring pupils from a primary school to a middle school at the age of nine. Such material was very useful to those schools which had taken part in the discussion, though perhaps less useful to other schools where the teachers were unaware of the nuances and borderline decisions to include or exclude a topic.

6   There are problems of after-care which need to be solved. The dialogue has continued to some extent, especially when lecturers have other duties in the same area, but there is a danger that the impetus is lost when the regular support is removed.

At the time of writing (during 1979–1980) only one lecturer is continuing this work. The other two lecturers hope to return later. It has been a thoroughly worthwhile exercise. It is difficult to assess the benefit gained, for often it is a slight change of attitude or a little more confidence, which builds up later into a noticeable improvement. Clearly there were some teachers who gained nothing; it is impossible to educate those who will not listen. But the fact that it worked so well is a tribute to the diplomatic and sympathetic persistence of three talented lecturers, who knew when to keep quiet, when to offer advice, and were willing to give much more than the nominal half-day per week when necessary, and to those teachers who accepted the challenge of a new and fruitful partnership with vigour and enthusiasm. It has proved to be a particularly effective use of college of education lecturers. Some of my colleagues would go as far as to claim that it has been a 'best buy'. [O]

*   *   *

In retrospect, perhaps the word 'catalyst' used in the heading of this section was not appropriate. In a chemical reaction, the catalyst itself is unaffected by the changes which it assists. Would these individuals say the same of their own roles?

Professional development is not only for teachers in school. It is no less important for those whose work lies in the training of mathematics teachers.

CHAPTER 6

THE ROLE OF THE LEA

In England and Wales the unit of educational administration is the local education authority (LEA), which covers the area of a county or a large city, and is controlled by an elected local council. The LEA employs the teachers, and equips and services the schools; subject to central government approval, it decides what system of schooling it will have and builds the schools; and it is legally responsible for the curriculum – although, in practice, detailed decisions are delegated to the headteachers of the schools, and by them to heads of departments.

There was a time when an LEA could put up schools, put desks and teachers in them, pay the teachers once a month, and consider that it had done its duty. Not now. The large sums of money spent on the education service have made authorities more conscious of their accountability to the public which finances them. Curriculum innovations such as computing call for expensive equipment, which schools expect the LEA to provide and maintain; in return, the LEA expects to exercise some control over the way in which it is used. Large-scale curriculum development needs support from outside the school for its successful implementation; most LEAs have a strong advisory service for this purpose. (The various titles used – 'inspec-

tor', 'adviser', 'advisory teacher', 'curriculum development officer', etc. –
indicate subtle differences in the interpretation of the role.) Most present-
day LEAs regard the professional development of the teachers they employ
as a major responsibility. This chapter describes some ways in which they
make provision for it.

Courses

For many people 'in-service training' means 'going on a course'. Certainly
this was where in-service training began as a recognized activity, and the
model is found in many walks of life – industry, nursing, the army, etc. –
besides education. It has become fashionable to play down the 'instruc-
tional' course; but we ought not to forget that it can still offer an important
source of intellectual stimulus and personal contacts for the classroom
teacher. The first vignette is by a curriculum development officer from a
city LEA; it gives some indication of the range of mathematics courses
offered in the LEA, and of their concentration in primary mathematics. In
general, secondary teachers are less avid course-goers than are their primary
colleagues; they also tend to draw on a larger range of providers, so that
many of their activities are described in the next chapter.

A curriculum development leader

The provision of mathematics courses varies considerably from one local
education authority to another. My own experience has fortunately been
with an authority which has been prepared to provide substantial sums of
money in order to refresh its teaching staff, and also to give schools concrete
support in terms of mathematical education. A few authorities, however,
provide no money at all; they make no provision for outside speakers or for
their teachers' expenses.

Late in 1976 Mr. Callaghan, then Prime Minister, made a speech at a
conference in Oxford in which he referred to the need to examine standards
of mathematics in school. This speech has aroused public interest in
mathematical education, and one result has been a large increase in demand
for in-service education, particularly in mathematics.

Let me describe the courses provided by my own authority (a city with a
population of 500,000) for just one year (1979–1980). These were many and
varied, from a course for infant teachers of six half-days to a course in eight
evening sessions discussing the mathematics curriculum for children aged
eleven–sixteen. In all there were four courses for infant teachers, four for

junior teachers each lasting for between six and eight half-days, one for
teachers holding posts of responsibility in primary schools, one on computing at secondary level, one (of eight half-days) for secondary probationary
teachers, and one for experienced teachers on the mathematics curriculum.
In addition, at secondary level six meetings a year are arranged for the heads
of mathematics departments in the schools.

Here is a selection of course programmes:

Infant teachers (8 sessions)

1   Sorting and matching
2   Ordering; cardinal and ordinal number
3   Number for middle infants
4   Number and money for top infants
5   Planning and organization
6   Recordkeeping and assessment
7   Using a thematic approach
8   Some aspects of measurement

(Course providers: 3 headteachers, 3 deputy heads, and 1 class teacher)

Junior teachers (9 sessions)

1   The mathematics curriculum
2   Infant/junior transition
3   Using structured apparatus
4   Addition and subtraction activities
5   Multiplication and division activities
6   Fractions and decimals
7   Using a geoboard to advantage
8   Logic and sets for juniors
9   Graphical work – a progression

(Course providers: 1 curriculum development leader, 2 class teachers,
                   and 2 college lecturers)

Primary teachers in charge of mathematics in their schools (9 sessions)

1   The mathematics curriculum
2   Aims and objectives
3   Why teach logic?
4   Standardization of methods
5   What about the gifted child?

6 Scheming a topic
7 Resources and support for other colleagues
8 Applying mathematics skills in science
9 Assessment

(Course providers: 1 HMI, 1 curriculum development leader, 2 college
               lecturers, and 1 LEA support service coordinator)

Secondary probationary teachers (6 sessions)
1 What are the problem areas?
2 Basic number work
3 Fractions and decimals
4 Understanding and motivating the slow-learning child
5 Probability
6 Mathematical games and other purposeful activities

In this last course four of the sessions grew directly from the participants'
'demands' during the first sessions – although from a planning viewpoint
some of this was anticipated!

(Course providers: 2 curriulum development leaders (for mathematics
               and special education), 1 college lecturer, and 1 class
               teacher)

A large variety of personnel is used to run these courses (a total of over
eighty sessions). These include as many teachers as possible who are already
using 'good practice' within their own schools. We also make much use of
lecturers from the local college of higher education. Approximately 25% to
35% of the 'lecturing' would be undertaken by myself (the curriculum
development leader for mathematics) and one other senior colleague who is
also a mathematician.

Some 'courses' are organized with discussion as the means of communication; many are practical and of a 'workshop' nature. All courses enable
teachers from different schools to meet and discuss the logistics of their
different situations. Interaction between course providers and between
course participants is encouraged.

Courses are evenly distributed with regard to 'school time' and 'out-of-
school time'. The local education authority attempts to provide as much
in-service work as possible 'in the firm's time'. One of the consequences,
however, is that for courses in school time only one person per school
usually attends, which makes for lack of dissemination or discussion back in

the staff room at school. Out-of-school courses, on the other hand, can and often do attract several teachers from one school, and there is evidence to suggest that discussions and ideas from the course will then be further explored at school.

In addition to these courses, some schools ask for courses for their own schools. This type of school-based course is proving popular, although very time-consuming. Nonetheless it offers the advantage that the whole staff of one school is involved. [P]

Teachers' centres

The first teachers' centres were established in England in the early 1960s, and they proved so successful that they are now numbered in hundreds – many in the centres of large towns, but some also in rural areas. Most are 'general centres', available to support all areas of the curriculum (including, of course, mathematics); but some LEAs have set up specialist mathematics centres, one of which is described below.

Typically a teachers' centre is a building with a complex of rooms such as:

a large room for lectures and workshop activities;
a sitting room for informal discussions;
a library;
a room for storage and display of equipment;
a kitchen;
a printing room, etc.

It will be under the control of a 'warden' (who will be an experienced teacher) with his assistant and secretarial staff, and on most days of the week it may be open from 9 am to about 8 pm (and even later when there is some special activity or meeting). It is the place to which teachers naturally come for advice, to look at equipment, to attend courses and meetings.

Running a mathematics centre, by a warden

Perhaps a more truthful title for this article would be 'Running my particular mathematics centre', for it is my experience, based on the last ten years as warden of one mathematics centre and as senior advisory teacher for mathematics, that will form the major part of this contribution.

In almost the same way as one hopes that the work of the mathematics centre is reflected in the schools, the hopes, desires, and needs of the teachers should be reflected in the centre. I say 'almost', because in many cases they need to be magnified. How do you judge whether a centre is a success or not? This is probably as thorny a problem as that of trying to decide whether a particular teacher is a success or not. Low or high numbers of teachers attending the centre, as with low or high numbers of grade 1 examination passes, give no true measure of the value or worth in either case. However – and taken to extremes – if the attendance is exceptionally low, then the centre is open to the criticism that it is nothing more than a very expensive office. Conversely, if the centre facilities are in full use, there is an excellent chance that the schools will receive some benefit. Perhaps here we have arrived at our answer; a centre should be judged to be a success or failure on the response to the question, 'What is going on in the schools as a result of the centre being in existence?' To affect the work in schools we must involve the teachers, and the way to attract the teachers is simply to ensure that the centre does cater for their needs. Having launched two general centres, I know how difficult it is to cater for every subject that teachers may require; but with a specialist centre such as ours, the teachers know they will see mathematics in depth and should stand a very good chance of finding the answers to any questions or problems they may have. A general centre, because of the enormous variety of subjects, can only deal superficially with each area of the curriculum. One advantage of a specialist centre – the warden might say disadvantage – is that the staff not only devise and arrange the courses that will take place, they also act as the instructors, lecturers, or leaders for these courses.

The needs of teachers, of course, are vast. But if I leave aside what I would term the clerical side of teaching – duplicating, photocopying, etc. – it is my experience that teachers want to know what to teach, how to teach it, why, and when; and a most important point to bear in mind is that they need the time to seek the answers to their questions. Having the dual responsibilities of senior advisory teacher and warden gives me a distinct advantage in seeking out, identifying, and providing for the needs of teachers. In my advisory capacity I have free access to all schools, classes, and teachers in the local education authority, and so gain insight to teachers' needs from first-hand knowledge and experience. Visiting teachers on their home ground, and talking to them individually, is central to the whole of my work and to the success of the centre. Having assessed the needs, wearing my advisory hat, I then change to my warden's hat and use the mathematics centre's resources to meet those needs.

At this point it is worth mentioning that access to a separate, limited sum of money, which can be used to support schools which wish to embark on a new course of action, is of great benefit and acts as an insurance policy against waste of new-found knowledge. In my view it is a pointless exercise for teachers to attend a course at the mathematics centre that involves the use of some structural apparatus, only to be denied the opportunity to use their new-found knowledge when they arrive back in their own schools because the schools cannot afford to buy the necessary equipment. In these circumstances I am able to purchase the equipment for them, or, if the particular equipment is specialized and expensive, provide it from our loan service for a limited period.

About nine years ago it occurred to me that if lecturers at colleges of arts and technology, teacher training colleges, and universities have time set aside during the week for them to carry out research and to keep up-to-date with their subjects, then teachers who teach seven or eight periods a day every day need just as much time for research and to keep abreast of modern developments. The problem was, and still is, that there is not enough money to provide substitute teachers to take the classes for teachers wishing to attend the centre. So I devised a mathematical 'obstacle course', which consisted of a series of unrelated mathematical activities for children aged roughly from nine to eleven, to which the teacher could bring his whole class. While the children, working in pairs, are gainfully employed, the teacher is free to look at apparatus or books, discuss ideas or – as happens in most cases – work with the children using different pieces of apparatus and in situations that are only starters to what could well be whole courses for teachers in the future. So much worthwhile discussion has taken place, and so many examples of what children can and cannot do have been experienced, that this activity has proved to be very popular with the teachers; so much so that 99% of the centre's activities are arranged purely at the request of the teachers, and bookings for activities need to be made a term in advance.

Perhaps many of our attempts to teach teachers in the past without children being present were rather like trying to teach carpentry without the wood. Of course, our activities are not solely restricted to work with children. We do run formal courses for teachers, which may take the form of a ten-week course on number designed to show *A way through number* – when sets are used, why they are used at a certain stage, where and when logic blocks should be used, and why they should be used in the build-up of number awareness. Also some insight is given into the usefulness of this work at the secondary stage. In other words, we try to give an overall picture of the subject. At the same time we remain very aware of the demands made on teachers' time, particularly primary teachers, who could be attending courses on different subjects every night of the week, and who therefore do not want to sit and listen to too much theory all the time. They want something to do in the classroom tomorrow morning, and so we attempt to make our sessions as practical as possible, often with run-off sheets that can be taken back to the schools for use the next day.

No mathematics centre would be complete without an extensive and up-to-date library of text and reference books; but alongside this should be a permanent display of mathematical teaching apparatus. The cost-effectiveness of such a collection is now beyond question. To look at a piece of apparatus displayed in a catalogue is one thing; actually to handle it and to talk to other teachers who have used it is quite a different matter. We have items at the centre, that may have cost us £27, which are, in fact, useless; but by drawing the attention of only ten schools which might have been considering buying this piece of apparatus to this fact, we would have saved the authority £243 on one item alone.

We have many sessions each term with parent-teacher associations, young farmers' clubs, students from colleges of education, institutes of education, the university department of education, visitors from other areas, and visitors from all over the world. Once each term an evening is set aside for what we call our Secondary Mathematics Teachers' Group. The first part of the evening is purely a social affair, when the teachers meet each other and discuss common problems over a drink. I must say that the size of the bar seems to increase term by term. The second part of the evening is devoted to some topic of general interest, and we usually invite an outside speaker or well-known figure to address the meeting. The record of attendances shows that these functions are extremely popular.

There are schools with classrooms where children go for 'mathematics lessons'. There are schools with classrooms where children go 'to do some mathematics'. There is a big difference between the two. I believe that a mathematics centre should be like the second school. It should have an exciting atmosphere, so that even the most doubtful visitor is persuaded, on entry, to pick up something and to start doing some mathematics.

The standard of work produced by a teacher is often a reflection of the importance someone else has been seen to attach to his work. Providing a mathematics centre for the mathematics teachers may be all that is needed to maintain active and enthusiastic participation in the development of the teaching of mathematics. [Q]

Curriculum development

When we first began to hear about 'curriculum development', it was assumed that this was a process carried out by a group of 'experts', who produced 'materials' and who then ran 'courses' to tell teachers how to use them.

It did not take long to learn that the relationship between 'curriculum development' and 'professional development' is more subtle than this – that, indeed, involvement with curriculum development is one of the most effective means of professional development. Curriculum development began to migrate from the ivory tower to the local teachers' centre. Teachers took the output of national projects, worked on them, adapted them to suit local conditions, and eventually turned them into local projects. An important breakthrough was the Schools Council's *Mathematics for the Majority Continuation Project*, a nationally funded project whose materials were developed in a number of such local groups. Other important examples are the Kent Mathematics Project, SMILE (a programme of individualized learning developed by London teachers), and the Birmingham Structured Mathematics Scheme; but there have been many others on a less ambitious scale. Clearly the local education authority, by providing advisory and resource support for a group of teachers in its employ, is well placed to act as an 'enabling agency' for this kind of curriculum development.

This chapter closes with descriptions of two very different pieces of local curriculum development which have come about in this way.

A head of department in a secondary school

'A supervisor has asked for a spot check. Does this box contain 1000 resistors?' Faced with these instructions, the two students started by counting out the resistors one by one, but they soon gave up this tedious method. 'There must be an easier way,' they remarked together. This is when they noticed the balance on the bench in front of them, and they were soon devising a system of counting by repeatedly balancing one pan of resistors against another.

This is one of the tasks which comprise the Cambridge Secondary Relevant Mathematics Project. The idea for the project had occurred to the warden of the mathematics centre several years ago. In September 1978, in response to the plea that the mathematics taught in secondary schools should be seen to be relevant to work and everyday life, the time seemed ripe for making the idea a reality. He discussed the idea with me; I must have appeared enthusiastic, and the project got off the ground.

The aims and objectives for the project are as follows:

TO DEMONSTRATE, through the use of a series of practical real-life situations, the relevance of basic school mathematics to life after school.

TO PROVIDE motivation. As all of the tasks are taken directly from everyday situations and have not been concocted simply for the sake of utilizing some aspect of school mathematics, the motivational effect is extremely great.

TO SHOW the need for accuracy. Each task requires the pupils to analyse a situation, solve the problem, and do some calculations, but all the time knowing that once their calculations are finished the task must then be completed practically, based directly on the findings of their calculations. The proof of the pudding is in the eating. The look of embarrassment that dawns on the pupils' faces, when 30 m^3 of ready-mixed concrete is pouring into their foundations, when in fact only 3 m^3 would have been sufficient, is a sight worth seeing. Perhaps for the first time in their lives they realize that 6, 7, 8, or even 9 out of 10 just won't do in a real-life situation; they must get 10 out of 10.

TO ALERT teachers to the fact that applying school mathematics in real-life situations requires much more from the student than the simple mathematical techniques practised in class.

The project was launched in an embryonic form in October 1979 as part of the Cambridge mathematics fortnight. Prior to this we had spent many months slowly building up and refining the relevant materials and tasks, a process we can see continuing indefinitely.

Teachers were invited to bring groups of fifteen- and sixteen-year-old students to the mathematics centre to take part, in pairs, in the project. The tasks at that time numbered about eighteen; we have since added to this total. Here are some examples of the tasks they were faced with:

Devising a method for cutting a tile so that it accurately fitted a given gap – then actually doing the cutting and fitting.

Using a variety of commercial timetables to find the time for a journey from Cambridge to Calais.

Taking the necessary measurements from a (nonrectangular) window frame needing to be glazed – then cutting a piece of card to fit.

Calculating the angle a lathe needs to be set to, to cut a taper of certain (given) dimensions – then measuring up a taper made by an apprentice to check the accuracy of the work.

Weighing letters and parcels and using Post Office leaflets to find the correct rates of postage.

Some of these tasks may seem simple and easy to carry out, but I would make two comments. First, that if the tasks are simple then that is how they really are. We have chosen the real-life situations according to three criteria: they must be real; their solving must depend upon some mathematics (or at least some arithmetic); they must fall within our budget and be portable, so that they can be set up at the mathematics centre or in a school classroom. But we have not chosen the tasks for their simplicity or difficulty, nor have we altered their level of difficulty.

Second, if the task seems simple – try it. You will soon find there is much more to it than first meets the eye.

This second point was forcefully demonstrated when we had a group of seventy or so secondary mathematics teachers in the mathematics centre working on the tasks one evening, and in many cases failing miserably to complete tasks which they had, at first glance, rejected as being too simple. One teacher told me he had 'no idea' how many notes and coins to ask for in order to make up some wage packets. Another claimed that it would not be worth the £10 installation fee to have a different kind of electricity meter fitted which gave a £7 saving on a (quarterly) bill.

Groups of students have been coming to the mathematics centre, with their teachers, for several months now and the project has been a great success. Students and in many cases also their teachers see, perhaps for the first time, that mathematics is needed in many situations outside the classroom and that applying mathematics to real-life situations is not as easy as it first appears.

A by-product of the project – but one we expected – is that the students enjoy taking part. In fact they are often reluctant to leave when time is up, and want to complete the task they are engaged on.

The tasks now number about twenty-five, plenty for the groups of thirty to forty we cater for. We are now considering splitting the project into three and developing about twenty or so tasks for each. These would be (i) tasks that everyone needs to perform – such as deciphering timetables; (ii) tasks students may need to perform when they go out to work – such as repricing dresses after a 12½% sale reduction; and (iii) tasks for students on examination courses which will give practical reinforcement to work done in the classroom, such as finding the centre of a rod of circular cross-section. [R]

*   *   *

A curriculum development leader

The group was set up in the mathematics centre in 1977 to produce material which would supplement a mathematics guide on number for the primary school, published the previous year. The decision to set up the group was a joint one between myself (the mathematics centre leader) and the local education authority inspector for mathematics. An added reason for such a group was a need to involve teachers from one area of the city which felt itself geographically isolated.

Several teachers already known to me were invited to the initial meeting, together with various nominees from local schools, about ten people in all. At the initial meeting the objectives of the group were discussed, together with a plan of the proposed outline. Decisions were taken about timing and frequency of meetings, and members departed to produce some initial ideas from different parts of the outline. It was decided to work towards producing a book on number games. Meetings would be held fortnightly, partly in school time, partly in the teachers' own time.

The initial work of each group member was eventually produced, and then copies were made to allow each person to read and digest all the submissions. Needless to say, each submission had a different style, and the first problem was to agree on what format to adopt. The next step was a major one: the first submission (developmentally) had to be pulled to pieces and eventually rebuilt. This sensitive operation needs handling with tact. Perhaps if the leader can put his head on the block first of all, then others will follow meekly! In other working groups, on occasions, the submission of a member of the group has been unacceptable. This again is a problem where diplomacy of the highest order is required. I leave you to your own salvation!

Our experience over several years has indicated that groups will not function unless there is a leader or coordinator who will be responsible for administration. This includes duplication of material for discussion and of draft writing as it emerges; it requires access to good reprographic facilities, especially some form of graphic design.

The following sections had emerged from our outline:

1   Infant number games
2   Place value activities
3   Number board activities
4   Bingo
5   Dominoes
6   Cross number puzzles
7   Magic squares
8   Number grid activities
9   Arithmogons

Slowly each section of the outline was tackled by the whole group; meeting by meeting fresh copy was produced and the draft edition began to emerge. As we progressed, minor alterations or additions to previous work were made, especially when cross-referencing occurred. When the final part of the outline was completed in draft form, all members were asked to read through the draft, bearing in mind style, presentation, format, and content.

Usually, when working with groups on these lines, I have approached headteachers to ask for a complete day or half-day release from school for a final revision of the draft copy. This, I believe, is necessary in order to focus complete attention on the document. Fortnightly meetings are too short to tackle a major read-through and revision; it takes half an hour at least to recall the previous thoughts and decisions on the work in hand, which leaves little more than an hour for producing new work. Also attendance is not consistent, since usually the group is made up of committed teachers with many other school responsibilities, which often (and rightly) take preference to our fortnightly meetings.

The draft copy, now revised with every 'i' dotted and 't' crossed, was then sent to the printers together with much discussion on layout, graphics, title page, etc. The final title, after discussion within the group and with other senior colleagues, was *Enjoying number work*. The idea of including the word 'games' in the title was considered dubious, as some people might reject the material on the grounds of children 'playing' in school! Titles can often be misleading, and certainly much thought went into this one before the eventual title emerged.

Next the proof-reading by two or three people working independently was completed, and printing commenced. Finally, a series of meetings was organized to launch the booklet, and on this occasion resource sheets were made available in four development centres spread around the city.

Evaluation of the exercise is our next step. No doubt the group members have benefited considerably – a valuable part of their own staff development. Our hope is also that the document will influence the teaching of mathematics in the classroom across a wider spectrum. [S]

CHAPTER 7

PROVISION OF COURSES
OUTSIDE THE LEA

The traditional method of in-service education is teachers' courses. These
are of a rich variety – long and short, full-time and part-time, some which
lead to a recognized qualification and others which do not, some in
mathematics alone and others in which mathematics is only a part. Sometimes they take place in school time, and sometimes they require teachers to
attend in the evening or in the school holidays. Some courses are provided
by LEAs, and deal with the specific needs which they perceive their
teachers to have. These were dealt with in the last chapter. Other courses
are run by universities, colleges of education, and other institutions of
higher education, and by Her Majesty's Inspectors of Schools (HMI). This
chapter is contributed by teachers who have attended such non-LEA
courses, and by lecturers who provide the courses.

Advanced courses in universities

Our first writer in this chapter describes her need for mid-career professional refreshment. She was fortunate; an able teacher, and well supported
by her LEA, she was seconded for one year of full-time study, to follow a

course for an Advanced Diploma in Education at her local institute of education. The second piece is by the mathematics tutor of the course which she attended, and discusses the need of many teachers to undertake further, specifically mathematical, study at some time during their careers. The third writer describes the range of work undertaken by the tutors in a large university school of education, and the contribution it makes to the professional development of teachers in its area.

The first ten years, by a secondary school teacher

At eighteen I left school vowing that, whatever else I did with my life, I was *not* going to teach. At thirty-six, having spent nine years in industry and having followed the progress of my own three children through primary school, I was overjoyed to hear that I had been accepted as a mature student by a college of education.

Four years later I emerged with a B.Ed. degree in mathematics, and what I thought was a sound philosophy for the years ahead. However, I was vaguely aware that stored at the back of my mind were some questions that scarcely seemed to be questions – and, if indeed they were, not questions that needed answering. For example, 'It matters what you think mathematics *is*, because that will affect the way you teach it'.

My first appointment was in a primary school, and during my second year of teaching it was decidedly 'mathematics across the curriculum' for all my pupils – we rotated parts of our bodies through a variety of angles in physical education; we scaled quantities for different numbers of children when we made sweets as part of our project on 'sugar'; we made a model of a city skyline in art and craft which involved constructing cuboids of differing dimensions. Creative writing also provided many opportunities for mathematical discussion – 'oranges' gave rise to fractions, 'bubbles' produced spheres, 'autumn leaves' led into area. The specimens collected on our nature rambles were arranged and rearranged in sets according to different criteria. Our science project 'water' afforded numerous opportunities.

During this period I found myself seeking connections with mathematicians outside the school, and decided that I needed to teach even more mathematics. I gained promotion in moving to a girls' grammar school and for the next two years was absorbed in exploring all the possibilities the syllabus offered. I was able to extend into the sixth-form under a very able head of department.

When the grammar school was reorganized into a comprehensive school the pastoral side demanded more thought, more staff were appointed, and 'politics' entered the staff room in no uncertain manner. It seemed that a focus for my teaching was needed each year and I became involved in 'links with industry'. This started with a series of meetings at the local technical college, attended by teachers, employers, and representatives from the field of further education, in an attempt to establish 'common ground'. This led to serving on a committee set up by the Secondary Mathematics Panel to investigate the possibility of a core mathematics syllabus for schools in our region. It also led to my undertaking some careers work within my school.

Mixed-ability teaching was being encouraged at this time, and I was well suited to this, in view of my primary experience. The children who needed remedial help brought to light another area that demanded exploration and, through a workshop at our mathematics centre, teachers from several schools pooled ideas for games, activities, and a possible curriculum for these children.

Meanwhile a hierarchical structure emerged in the organization of my school and I felt excluded from the decision-making; in fact I had become aware over the years of a gradual erosion of my autonomy since the days when I worked in the primary school. The answer seemed to lie in seeking promotion, or perhaps a change of school, but as I started to write letters of application and to be interviewed, I became aware of those questions which had not seemed questions at the time. My philosophy no longer seemed adequate. The 'back to basics' movement seemed justified and yet pupils reacted with 'This is not the mathematics you used to teach us – can't we do more of that sort?' Where was *I* going? What *did* mathematics mean to me? I applied for a year away from school to study further.

The process of self-examination is a painful one – depression sets in when one realizes how limited one's own teaching has become, how 'out-of-date' one is. The Advanced Diploma is not about gaining another qualification, and therein lies its value. I have chosen to write essays on topics that I feel uncomfortable about, such as the demand for a statement of one's aims (fair enough), but specific objectives somehow seemed a challenge to my professional status. This is a new attitude to study, in contrast to my work in college where I tended to choose those themes that I felt I could do well.

Meeting other subject specialists in a structured situation has made me appreciate their ways of looking at life in a manner that is not possible in the informal atmosphere of the staff room. I have also become aware of areas (such as perspective drawing) that are no longer covered for all pupils by the formal curriculum.

As we visit other schools in session ideas are stored – indeed, I am very like a squirrel hoarding nuts for the next ten years.

In these days of 'accountability', I am very conscious of making the most of this opportunity. I believe that, leaving aside the question of promotion, which in any case is likely to become increasingly more difficult in the profession as a whole, I shall return to my school with a new perspective, a wider view, renewed in spirit, and a better teacher. I shall also have acquired many new skills, particularly those in the use of the computer.

It is my belief that *all* teachers require such a period of refreshment if they are to maximize their potential – a very necessary prerequisite if they, in their turn, are to maximize that of their pupils. [GG]

*   *   *

The tutor's view of the Advanced Diploma

Mary had been teaching for about fifteen years. She had a good degree in mathematics, and regularly attended local meetings of the Mathematical Association. But she felt she was getting stale. Her school had been reorganized and had lost its academically challenging pupils; the headteacher gave little support, believing that mathematics was an unimportant subject for girls. What Mary needed was simply refreshment and encouragement – the chance to read and do some mathematics for herself, to attend some stimulating lectures, to enjoy some mathematical conversations.

Dick had taken a university course in mathematics, but admits that it did not mean much to him at the time. The lectures were generally formal and algebraic, and had done nothing to develop his geometrical thinking; when he started teaching, that is how he taught. He found it tough going learning about transformations and groups of symmetries – but he reckons he's a much more lively teacher as a result.

It is not often that primary school teachers want to make a special study of mathematics, but Karen was an exception. She had enjoyed her mathematics at school and college, and wanted to learn some more. But she had one 'blind spot', and that was probability. Three or four times she had attended courses on it, but made little headway. So we talked about some of the probability experiments that one might do with eleven-year-olds. What would we hope they would learn from them? How many times would we need to repeat an experiment for the pattern to emerge? It was not long before Karen was confidently discussing binomial probability, chi-squared, etc. – but in a context she could understand.

Dan had taken a course in statistics at university. He knew how to use the

Poisson distribution and how to apply a t-test – but he had no idea why. So we approximated to them by situations based on discrete models; we compared them with the results of random experiments, using the computer as a tool; finally, we examined the possibility of developing them by mathematical analysis. Hopefully, even if he only teaches his pupils 'how to use them', he will do so with more insight.

Joan had been a good mathematician at school, but – although she got a moderately good degree – she had never really caught on to the subject at university. She had a few years away from teaching to bring up a family, and was now doing a competent enough job as a secondary school teacher; but she didn't really believe in herself as a mathematician. So for her the course offered a second chance: she got really involved in the symmetry group of the triangular antiprism, wrote a batch of computer programs to help her in teaching sixth-form statistics, and wrestled with the foundations of affine geometry. The last time I heard from her she was thoroughly enjoying running the mathematics department in her school.

Mary, Dick, Karen, Dan, Joan . . . were all first-class teachers with several years' experience. Otherwise they would not have been selected for 'secondment' (absence on full salary) for further study. It is sometimes asked why they need to learn any more mathematics – after all, they have all had three or four years postschool education; should this not be enough?

From my experience the answer is 'no'; and the reasons vary from one teacher to another. Some older teachers are being asked to teach in school topics which they never encountered as students: groups, linear algebra, statistics are examples; and many teachers have never used a computer. Others were too immature at university to appreciate that there is more to mathematics than a string of theorems and techniques to be learned, and they need to ask *why* mathematics developed in the way that it did. For another group the everyday pressures of school life – sport, clubs, marking, discipline, administration – leave them with no time or energy to think about mathematics beyond the immediate demands of their teaching programme.

Anyway, it is not easy to keep up one's mathematics by oneself. First there is the problem of choosing a suitable book at the right level – challenging yet not too difficult. Too few books provide 'sign posts' indicating what is important and what can be by-passed; often theorem follows theorem at a monotonously even pitch, communicating little insight in the process. Then, where does one turn when one gets stuck? And if and when one reaches the end, will one be a better teacher as a result?

The course which Mary, Dick, and the others attended lasted for a full year, and half the time could be devoted to mathematical studies (the rest being allocated to some other area in the theory or practice of education). This allowed about 200 hours of personal contact with a mathematics tutor – of which about 50 were used for seminars on mathematical education and visits to schools, the remaining 150 for mathematics itself – and a comparable amount of time for private study. How should they be filled?

Certainly the most important question is 'how?' rather than 'with what?'. Content by itself is of secondary importance. These teachers do not need, or want, the detail of a formal university course: they are unlikely to be better teachers for knowing, for example, that 'in the group of automorphisms of a group G the inner automorphisms are a normal subgroup'. There needs to be time and encouragement to explore, to dip into the literature (including half-remembered articles from back issues of journals), to talk about teaching implications, to look at a piece of mathematics from several points of view – without the constraints of a rigid syllabus. The spirit in which mathematics is approached, and the insights generated, are in this context far more important than the mastery of technical detail.

For my own course, the framework provided for this purpose consists of computing (which then services the mathematics of the whole course), group theory (a platform for geometry), themes from statistics and probability, linear algebra and analysis, and (for some) mathematical modelling (which involves some extended project work). The topics are selected more for their ability to enrich and extend school mathematics than for their intrinsic importance; and different teachers take very different things from them. We hope that they emerge from the course not just as better mathematicians, but also as more complete teachers. [T]

*　　*　　*

A school of education

The work of university schools and departments of education in England and Wales varies greatly from institution to institution and it is impossible to generalize about the variety of their work. I intend, therefore, to consider only my own university where the school of education, being one of the largest in the country, has a diversity of activity which may be taken as typical of many other institutions.

Altogether we have 919 students in the school of education in the present year, of whom 283 are graduates following courses of initial training for teaching mainly in secondary schools. The remainder are working for one of a number of named awards, some of which are designed specifically for students from overseas; these account for 113. The others are studying for diplomas, master's degrees, or doctorates, all of which are able to cater for the particular needs of teachers of mathematics. In the master's degrees, for instance, there is both a full-time and a part-time scheme. Among courses which are of specific relevance to mathematics are included mathematical education, the learning of science and mathematics, the development of intellectual skills, and computers in education. In addition there are many other more general courses in education theory. Master's degree students are required to follow three of these courses and to write a dissertation of 12,000 words. For mathematics, the dissertation has covered a wide range of subjects chosen to suit the needs of the individuals involved.

At a level below that of a master's degree, a number of diploma courses are provided.

At present there is one diploma course designed specifically for teachers who are teaching mathematics in the nine–thirteen age-range. Other diplomas, for general secondary education, for special education, and for administration, may contain an element of mathematical education and a short course in statistics.

Award-bearing courses for diplomas and master's degrees have become very popular in recent years but there has been a marked trend away from full-time to part-time attendance. Part-time diploma courses have been introduced only in the last few years; in the master's degree area numbers are now 167 part-time and 37 full-time students, whereas only five years ago there were 50 part-time and 50 full-time students.

There has also been a marked increase in the total number of students seeking named awards in education, although it is only recently that teachers who wish to study a teaching subject, and mathematics in particular, have started to come forward. Formerly the most popular courses were those involving general educational studies. It is difficult to give precise reasons for this trend, but it is possible that general studies are more suited to those who are seeking rapid promotion in an expanding educational service, whereas in the current declining situation wi h less opportunity for rapid promotion, teachers are more concerned to take courses which will help them to consolidate their skills in their present post.

As a direct response to the present situation we are introducing a new master's degree in 1981 which provides a study of both mathematics and mathematical education. This course is aimed particularly at young mathematics teachers who are seeking promotion within a mathematics department and is similar to courses offered in a few other institutions in the United Kingdom.

Mention has already been made of the graduates who are following courses of initial training for teaching. In most universities they form the main body of students studying in education faculties and we are no exception. As job opportunities decline, many of these students are beginning to show more interest in the idea of continuing education through in-service provision, and it has become easier recently to establish the idea of in-service activity as a natural part of a teacher's way of life. Certainly many more recently qualified teachers are returning to the university after a short period of teaching to acquire further qualifications and, as a result, the average age of students on higher degree courses has come down. It is a trend to be welcomed, and makes more sense of the view that preservice training can only be a beginning in a much longer period of professional development. Even so, it is clear that it is the better students who come back, and we are no nearer solving the problem of how to involve the less-good teacher in in-service work.

Courses leading to a named award are popular with teachers because the award is 'marketable', often in terms of promotion chances. Our mathematics tutors are also extensively involved in in-service work of a more general nature, which does not lead to a named award.

We mount a range of courses which vary from single lectures (to secondary school pupils or teachers or both), to one-term part-time management courses for heads of mathematics departments in secondary schools.

There are also many courses of six evenings in length. These may be designed for primary teachers of mathematics or aimed at secondary teachers interested in the process of learning mathematics. Another series of courses has been concerned with the interrelation between mathematics and biology in secondary schools. Less formally there are consultancy-type activities, particularly involving the use of micro-computers in schools and also supporting local projects organized in schools in the area. The university also provides one full-time teacher fellowship which is designed to allow a local teacher to spend a period of time working on a specific problem. Nevertheless, all these activities are in-service activities of a formal kind. In addition we have many informal links with teachers who work closely with the mathematics team in a wide variety of activities including teaching practice supervision, workshops with students, group-teaching seminars, and resource development. These informal in-service activities are of great importance because they are usually associated with very real professional development.

The extent of the involvement of a university in these activities depends on the number of mathematics tutors available. In Britain the norm is for a small number (one or two) of mathematics tutors to be appointed in any one school of education. In such circumstances the range of activities is bound to be limited. In addition, another important aspect of a tutor's work is that of research. Many higher degree schemes contain a research component or are taken completely by research and, because staff numbers are so limited, research for higher degrees may constitute the main research contribution of the department. [U]

The in-service B.Ed.

There are many nongraduate teachers in British schools. Their initial training consisted of a two-year or three-year course at a college of education. When, in the mid-1960s, Bachelor of Education degrees first became available to students at colleges of education, serving nongraduate teachers began to demand an opportunity to take the B.Ed. degree as an in-service course. The first contributor, an officer of the Council for National Academic Awards (CNAA) describes the structure and development of the in-service B.Ed. in mathematics. About half of the institutions which provide initial and in-service training for teachers give CNAA degrees; the other half confer university degrees. The second contributor is a primary teacher, whose B.Ed. in mathematics was hard-won against the sheer exhaustion of a part-time degree student in struggling with the conflicting demands of school and study.

An officer of the CNAA

The B.Ed. degree has been available as an award for serving teachers in England and Wales for barely a decade. In the first newsletter of the DES-funded research project 'Preliminary Evaluation of the Inservice B.Ed. degree', Norman Evans stated that the first official reference by the DES was in a letter in May 1969 which referred 'to arrangements for serving teachers to take the degree from September 1970 onwards and to the possibility of part-time courses in the future'. The initial provision was therefore in the form of a facility to admit to the final year of the existing preservice degree courses 'suitably qualified teachers who have five or more years' experience as a qualified teacher'. Thus, teachers were required to obtain release for a full year in order to join in the degree course with students who were completing their initial training.

This became the 'traditional' model for an in-service B.Ed. course. The full-time preservice courses, which themselves dated only from the mid-1960s, generally consisted of concurrent strands of educational studies, main subject studies, and professional studies. In the fourth and final year, the stage to which in-service students were to be admitted, professional studies played a diminished role, if indeed they existed at all. Thus the in-service nature of the course was, at that time, solely a description of the point in time at which it was taken – after some experience as a practising teacher; the courses had not been designed to meet the particular needs of practising teachers. These courses did, however, provide opportunities for teachers to engage in the further study of the main subjects they had pursued in their Certificate in Education courses five or more years before, and they would facilitate in-depth study of mathematics by those who were previously qualified in the subject. The course would not, however, necessarily reflect their current concerns as teachers of mathematics.

The first development from this position was the provision of part-time courses based on the 'traditional' model, i.e., courses extending over two, three, or four years of day release and/or evening attendance and incorporating the study of a main subject alongside educational and professional studies. Norman Evans reported in May 1978 that fifty-nine part-time courses were then available, but by that time a further trend was also perceptible – only a minority of courses still required the study of a main subject, whether they were offered on a full-time or a part-time basis.

The first in-service B.Ed. course approved by the Council for National Academic Awards was typical of the new 'professional' trend; there was no opportunity in this course – which was launched by a polytechnic in 1973 – for the detailed study of a main subject at the student's own level. Nevertheless, courses continued to be accepted by the council which did make explicit provision for subject development. In one such course – operated jointly by a polytechnic and a college of higher education – students could decide after a common foundation course on 'Education: Theory and Classroom Practice', whether to follow a programme in which the study of an academic subject, together with associated pedagogical studies, figured alongside the study of an educational discipline or one in which the latter was complemented by applied educational options. Mathematics and computer studies were both included in the subject options offered, but in the event, in the first five intakes they were each only run once. The extensive range of options available in such a course made the achievement of viable subject groups difficult in other than a very few of the academic subject options on offer. This is a major factor in the pervading economic climate in determining the range of provision made available in in-service courses.

A further major factor in planning in-service B.Ed. courses has been the commitment by the colleges to providing courses of immediate relevance to the situations in which teachers find themselves; in these courses, students' ongoing professional experience would be used as a resource for their college-based studies and this would, in turn, contribute to their professional enhancement. This had been the rationale for the first course described above, but commitment to such a philosophy need not necessarily lead to the exclusion of all subjects of the school curriculum from an in-service B.Ed. course. Thus, for example, the revised version of the second course described above removed the wide range of subject options but included a wide range of professionally oriented options which included 'The Teaching of Basic Mathematics'. This was described as 'relevant to those engaged in teaching pupils in the 5–13 age-range in school, or in lecturing to courses involving basic mathematics in colleges of further education'. It would help students to 'develop the knowledge and skills which will enable them to communicate this subject more effectively to their pupils'. It was specifically designed 'to meet the needs of non-mathematicians involved in teaching mathematics' and aimed to enhance professional skills 'through a deeper understanding of the nature of the subject, its development in school and the learning process involved'. This course has not been alone in including a professionally oriented mathematics option; another institution proposed a full-time option in mathematical education and in a third institution 'School Mathematics' is offered as an elective course. These institutions have had to tackle the difficult task of developing syllabuses which are appropriate for both primary and secondary teachers and those whose mathematical background, in terms of previous formal study of the subject, may be very disparate. The one thing that all these teachers have in common, however, is that they all have some current responsibility for teaching mathematics, whether in the context of the primary or first school, or of the specialist role of the secondary teacher, or of an intermediate position at the lower end of the secondary school or in the middle school. Such options have to balance economic viability with educational feasibility – if too many restrictions are placed on entry, groups of recruits will be too small to warrant offering mathematics education in the B.Ed.; on the other hand if the intake is not restricted in some way the very heterogeneity of the group will jeopardize the worthwhileness of the course.

A further problem in such an option is whether the mathematical understanding of students should be assessed, or indeed whether some part of the option should be explicitly devoted to the development of the students' understanding of mathematical concepts and command of mathematical techniques. Options of this form usually attempt to balance both aspects; study of the nature of mathematics to enhance students' understanding of the subject complements the study of the learning and teaching of mathematics, although in practice an integrated approach may be adopted. Similarly, in assessment, students are required to demonstrate a level of mathematical competence and understanding appropriate to the level of teaching in which they are engaged.

The present position of mathematics in the in-service B.Ed. is by no means uniform across the country. Some institutions still offer an in-service course including the 'traditional' main subject; others may include options in mathematical education within a professionally focused course; whereas a further group may include no explicit provision for students to opt to develop their understanding of the subject and its pedagogy. Many courses do, however, include projects or special studies, particularly at the honours degree level, in which students may develop issues of particular interest to themselves. Even in a course without specific components devoted to mathematics or mathematical education, a project or special study provides an opportunity for teachers with relevant interests to investigate problems related to the teaching of the subject and building on the curriculum studies or psychological components of the course.

One university has developed a scheme in which one component of the degree – approximating to the main subject element – may be taken by an independent study mode. This enables teachers to study in depth areas of their own choosing derived from their professional involvement, even though logistical constraints would prohibit the provision of a formally taught component in these fields. Tutorial supervision in such cases is critical but the potential of such an approach is considerable.

The future of the in-service B.Ed. is generally assumed not to be of long duration; the last entry to three-year Certificate in Education courses took place in 1979 and as the supply of sufficiently motivated nongraduate teachers dries up the courses themselves will disappear. However, even in some of the areas in which courses have been established for more than five years, sufficient numbers of candidates continue to come forward to warrant offering the courses, although in some cases economic constraints prevent release from teaching commitments which were possible in a more favourable climate, and courses therefore depend on a level of teacher motivation sufficient to maintain attendance in evenings, vacations, or weekends on top of a full teaching load.

Teachers have, of course, not been restricted to the in-service B.Ed. as a road to graduate status; many polytechnics and other colleges have offered part-time degree courses in mathematics, statistics, and/or computing for more than ten years. The Certificate in Education with mathematics as a main study was, from the beginning, conceived as an appropriate entry qualification for such courses, and teachers were seen as a significant potential clientele. The institutions offering such courses also initiated the Polymaths certificate, which is an alternative qualification for mature students wishing to enter degree courses in mathematics. This might provide a means by which a teacher who is not well grounded in mathematics could develop an adequate basis on which to tackle a part-time degree course in mathematics. The Open University provides a third route to a degree for the nongraduate teacher who may choose to mix education credits with the study of other subjects. Teachers form a significant proportion of the clientele for the mathematics courses offered by the Open University.

In summary, therefore, a number of different routes to a degree exist for a teacher with an interest in mathematics. The B.Ed. provides a variety of opportunities, but will generally require a student to consider a wide range of professional issues as well as mathematics, which indeed may not be included explicitly in some courses. In all these courses, however, teachers are able to follow courses of study which can be related to their professional role and enhance their performance in it. While one mathematics teacher at secondary level may prefer a course which makes explicit provision for the extension of his mathematical understanding as an underpinning for a continuing teaching commitment, another teacher at the same level may prefer a course in which the links between the subject and its pedagogy are made explicit; others, especially at primary level, may see the professional orientation of the subject study as the major priority. At present a spectrum of provision exists; it is important that it is effectively exploited in the interest of the teaching of mathematics in the schools. [V]

*   *   *

A *primary school teacher*

When I first applied to do the in-service B.Ed. I saw it purely as an academic qualification which might further my career prospects. I suspected that I might gain some value from it in terms of classroom practice, but expected an academic approach relating little to the practical problems of teaching.

In fact I have been greatly surprised and pleased by the approach I have been able to take. Having been given a choice of options helped; being wisely advised in my choice has been more helpful. The problem with any course is that you don't really know what you are letting yourself in for. I was fortunate enough to take units which were both interesting and practical in the sense that one could consider theory with a view to classroom activity. My interest was stimulated because of this – after all I teach because I enjoy doing it; I gain satisfaction from helping children develop skills. In retrospect it was because I had the opportunity to relate B.Ed. work to children that much of the theory began to take on a real meaning. Taking theory and 'testing it' seemed natural since I have the opportunity to test.

Perhaps of equal, if not greater, importance is the attitude and interest shown by the tutors. Never in the time that I have been doing the B.Ed. has anyone not had the time to discuss the problems and criticisms of a particular approach. It gives one the impression – and I think this is important – that I am really doing some good, that what I think is viewed as worth thinking about – that I am part of a professional body. I was able to say, for example, 'You know how some children do this; well I tried so and so, and it does not seem to work. What do you think?', and I had instant feedback. I suppose that what I have learned is that I should think *more* about information rather than absorb it and make rash decisions; that I should be able to defend decisions rationally, but be prepared to accept that these might equally well be criticized. I like to think that this 'practical' consideration of subject matter has not confined itself to educational matters. In other words, if I am doing a piece of work, statistics for example, I like to think that I can consider a particular theory in terms of its broader meaning within the field – how it fits in. This is something which never occurred to me before I did the B.Ed.

In truth, however, I have never found any of the units easy. The work is made more difficult by the shortage of time, and the difficulty of finding books. There's always the nagging doubt that one is not doing one's job properly: 'Should I be doing more preparation?' I like to think that, given the time, I would get better marks for my essays, prepare myself better for exams, and so on. In some ways the process feels like an endurance test; it consists of cramming work in when time is available. Exam pressures feel different, probably because one does not have time to think about them. Last year, for example, I did an exam straight after a week at a children's camp. I was extremely tired, but not worried to any great extent by the exam. I knew I should be worried, but I was too tired. Sitting the exam I had consciously to drag my concentration onto the matter in hand as things began to slip. I use this not as an excuse but as an example of a different type of pressure. I had never met this type of pressure and didn't know how to handle it.

Another difficulty born out of 'cramming' work in when time is available, is that continuity of work is continually broken. It means that one tends to lose the thread of one's train of thought. I find it time-wasting in the sense that I often have to reread sections and rethink them to get into the same frame of mind.

I think that what is certain is that without the moral support of one's family, the B.Ed. would be an almost impossible task. The stream of seemingly endless tasks – school, home, and B.Ed. – are burdens greatly lightened by encouragement.

Despite appearing to pour cold water over the idea of the in-service B.Ed., I think it has been an immensely valuable course for me. Being able to relate new learning to school situations is most important. The only suggestion which I would make, so that the opportunity for this type of study is not lost, whilst easing the burden of work, is that the course should be run over one or two days each week for three years. [W]

The Open University

The major provider of continuing education at degree level for adults is the Open University (OU)*. Since its inauguration in 1971, many teachers have graduated through its distance-learning programmes, which consist largely of correspondence courses, supplemented by radio and television broadcasts, teams of local tutors, and full-time summer schools during the summer holidays. For those who already have a degree, or who do not wish to undertake the full degree course, the Associate Student programme provides an opportunity to take individual courses, either parts of the undergraduate programme, or courses specifically designed for particular groups of people. For instance the Diploma in Reading Development was designed to develop on a sound theoretical foundation the teaching skills of serving teachers and it required regular classroom work as an integral part of the course. The first in-service course in mathematics for teachers mounted by the Open University is being offered for the first time in 1980. It is an attempt to break away from the individual nature of OU work – the student works for himself and tries to pass the course – to a cooperative school-focused model of OU in-service education. Teachers are required to do projects which, by their very nature, involve others in the school in discussion.

The following piece is written, alternately, by the chairman of the course team of 'Mathematics Across the Curriculum' and a junior school head-teacher who took part in the development work of this course. The development work was undertaken with the cooperation of a particular LEA and a group of its teachers – at least two teachers from each of the schools involved – and it is the OU's hope that other LEAs will take out block bookings for the course and use it in the same way.

Chairman of course team

In England and Wales a theme of popular attack on mathematics education has been on the failure of 'new maths' programmes to improve basic skills. It was against such a background that a group at the Open University started work on an in-service course for teachers called PME 233: 'Mathematics Across the Curriculum'. The group began by trying to iden-tify what constituted basic skills and, in particular, what rationale could be found for classifying one skill as basic, another as (let us say) a frill. A useful starting point proved to be the following quotation from the HMI document *Mathematics, Science and Modern Languages in Maintained Schools in England*.[9]

> The only justification for including mathematics as part of the compulsory curriculum for all children is the power it has to explain. The person who understands mathematics can understand other things more quickly. But very few people obtain this power by learning mathematical skills in isolation: unless most people see the applications of mathematics as they proceed they never come to see them at all. Skills should be developed in context.

The idea that 'skills should be developed in context' made it imperative that PME 233 should be directed towards improving the teaching of *using* mathematics rather than of teaching mathematics as a subject in its own right. This made two novel demands:

1　The justification for including a mathematical topic should be its usefulness (rather than the intrinsic demands of mathematics as a subject).

2   A context was needed in which all teachers (and hence their pupils) could immediately find applications for the basic skills they were teaching.

The approach to these two demands has been firstly to analyse the needs of adults for using mathematics, and secondly to develop the notion that children can, by their own action, solve real problems whose solution has some effect on their daily lives.

If the Third International Congress on Mathematical Education revealed anything, it showed a widespread world-wide dissatisfaction with the lack of success of curriculum development projects. The role of the teacher was at last formally recognized to be central in the learning process. Broadly speaking children nourished on a 'new maths' diet do not perform more or less well than those fed a more traditional fare. What remains is a feeling that the teacher needs help rather than just a new textbook. The specific help needed appears to be that of reestablishing, some would say establishing, in mathematics teachers the confidence that what they are teaching is important and relevant. The approach of PME 233 to this has two parts:

1   to develop in teachers a deep understanding of the ways in which mathematical thinking is used or could be used by adults in everyday life;
2   to recommend the process of 'real problem-solving' as being one way of developing a similar understanding in children.

This report focuses on the latter aspect and therefore it is worth considering in more detail what is meant by 'real problem-solving'. On the one hand, there are many adult situations where mathematics plays an important if not crucial part, while, on the other hand, the general diet of mathematical activity in school has very little in it of a real nature.

But this is not to say that there are not as many real problems in schools as the majority of adults will encounter in their work. Playtime, clubs, school meals, getting to and from school, these all present problems within the school whose solution has a great effect on children, as well as on teachers. The notion of reality also provides an educational objective. Children need to discover for themselves that they can, by their own action and by making their own decisions, confront and solve problems that have real meaning for them, and whose solution has some effect on their daily lives. Solving complex real problems is something that children are perfectly capable of, as is demonstrated by the work of an American project called 'Unified

Science and Mathematics in Elementary Schools'.[10] 'Real problem-solving' was seen, therefore, as providing a context in which teachers could help children to develop their mathematical skills, thereby answering some of the HMI criticisms.

Junior school headteacher

Our school is situated on a busy main road in an urban area. The site is abutted by a large factory, a railway line, and a canal. The buildings are 101 years old and have both the advantages and disadvantages of their years. The parents and grandparents of many of the children attended the school, and the community from which they come is closely knit.

We have much experience with children who have limited language development and problems with reading and writing. Likewise we have children who find equal difficulty in mastering the mathematical skills. We also have children who are both naturally talented and those who function well because of a caring home background where interest is shown and support given. We aim to provide the children with the experience and skills which they will need to enter into the community of which we are a part and to function within it. We also aim to give the children those skills and experiences which they may not be able to develop from within a narrow environment, and to widen their horizons to a world outside.

I feel that the day-to-day demands on a teacher in an area like ours call for a great deal of experience and expertise in areas outside teaching in the classroom. At the moment, education cuts are resulting in a reduction of the numbers of teachers, and yet the demands upon our time and resources are mounting. It is more difficult to release teachers during the day for in-service training and therefore we are engaged in 'in-school' courses after school hours, involving members of staff with expertise, as well as local education authority advisers. We are constantly assessing and evaluating the content and relevance of our curriculum and this too means that discussion and exchanges of views, experiences, and ideas must take place in frequent staff meetings.

Chairman of course team

Among the general criticisms of in-service work are the following:[11]

> Theoretical courses may be interesting but seldom achieve measurable change in the classroom. In-service work must be closely related to classroom practice and should stimulate classroom activity as well as providing a chance for reflection on what has been achieved.

The measure of the in-service credibility of PME 233 had to be its direct effect on classroom practice.

Thus it was the intention of the course team to devise a programme designed to focus on the teacher's role, rather than to introduce new materials for pupil use while ignoring the teacher's essential task as a manager and generator of innovation. The 'real problem-solving' section of the course was the main pedagogical device for doing this. It makes the following provision:

1   A two-day residential course serves to help the teacher to identify the problem involved in introducing change into his teaching situation.
2   The subsequent course units (5–8) help the teacher to further define, consider, and tackle this problem of actually implementing any new ideas in his own school situation.
3   Tutorial support assists the practical realization of the ideas in the classroom.
4   A project write-up enables the course team to examine the teacher's ability to translate course philosophy into practical experience.

It was clearly necessary to try out the PME 233 materials before they went into print. For this purpose, the course team ran a trial version of the 'real problem-solving' section of PME 233 in a way that would, as nearly as possible, reflect the experience that teachers would have. Thus a two-day residential course was run at a convenient site for a single LEA. Meanwhile, the advisory staff of the LEA agreed to provide thirty teachers to take the course, and to monitor and evaluate the effectiveness of the programme in the classroom.

Junior school headteacher

As our 'real problem' we chose to investigate the possibility of converting a walled piece of derelict ground within the school site into an attractive school garden. The problem itself was not only real to everybody in the school, but to us adults at times it seemed impossible to achieve. I had already sought professional advice as to the practicality of such an undertaking and had been advised to hire a mechanical digger. However, another teacher and I decided to discuss the possibilities of a garden with a mixed-ability class of second-year juniors. They were very excited and motivated because they could immediately envisage the finished garden with many embellishments and at first they disregarded the practicalities. We built a topic web and group discussions began. At once the children were using a number of skills that they already had and themselves saw the necessity to learn new ones. They were thinking, communicating, writing, and calculating with a purpose. The children visited local shops and looked through seed catalogues, asking for and receiving free samples. They used the telephone, an unpractised skill, which was rapidly learned and utilized with poise and purpose. They duplicated letters to parents inviting fathers to help us to dig the ground, and they reasoned that such labours must be rewarded, so they advertised the perks. They received a response and, as a result, one hot summer's evening the ground was dug.

Meanwhile, another group had planned the garden; they had measured the land and drawn a scale diagram and planned the shape. There was already part of a path laid and they soon calculated this was not sufficient for our needs, so they planned an extension using materials which lay around the site. A further group talked about the plants they would need, and designed an experiment to find which areas of the garden were in the shade and another to discover which areas received the most water when it rained. Unbelievably, the second experiment was unproductive; during the three weeks before the end of term not a single drop of rain fell. The first experiment, however, was successfully completed and involved careful measurements of shadows throughout the day; the result was shown on a graph.

Chairman of course team

After the two-day course, the participants went back to begin a 'real problem project' in their own classrooms. They had seen the value of real problem projects and were happy that the children would benefit in terms of investigative procedures, language development, and social development.

But there were two questions that only classroom work could answer:

1   Have we any real problems in our schools?
2   Where is the mathematics?

The teachers had seen evidence from two examples discussed at the course but were unsure as to whether the same positive results would apply in their own classrooms. What are the criteria for a 'real problem'? The first is that the problem should have immediate, practical effect on the children's lives. Secondly, the children must be able to effect some improvement of the situation. Thirdly, the problem should not have a known solution; we need to get away from the old syndrome of 'the teacher at the board playing a game of guess-what-I'm-thinking'. Fourthly, the problem should require the children to use their own ideas to solve it. Finally it should be big enough to involve the whole class.

The teachers soon reported that they had indeed found all sorts of problems, such as the following:

1   Tuck shop problems: not just whether the school should have a tuck shop or not, but how to work a tuck shop effectively and was it worthwhile?
2   Planning visits: it was the summer term and a time when a lot of visits were taking place. Could the children be involved, could they have a say in the school visit?
3   School gardens: could waste ground be utilized and could children help to solve the problem?

All sorts of problems of safety were identified, such as road safety and problems when children have to cross demolition sites and go near canals. One headteacher on the course went back to her fourth-year class and asked them to write down problems which they thought they could tackle. The class provided her with twenty-four problems, specifically related to their school.

It was not until the projects were well under way that the teachers began to see the potential for using mathematics. At the two-day course the idea had been introduced that two kinds of opportunities would arise. First there would be times when the children would be able to practise skills with which they were already confident. However, there might well be occasions when the children could make no further progress without learning a new technique. In addition, these two aspects would, in a sense, feed on each other as in the following diagram.

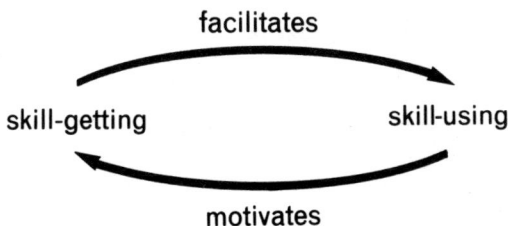

facilitates

skill-getting   skill-using

motivates

With this in mind the teachers made notes on the ways in which mathematics entered into their projects. The analyses for individual projects were then amalgamated to produce the data shown in the table below. As this table indicates, a range of mathematical ideas arose naturally as the children worked to structure their experience with purpose. Furthermore, it was not only a question of consolidating and reinforcing skills already learned. The children needed at times to acquire new skills in order to achieve their objectives. Certainly the teachers agreed that they had managed to answer the question 'Where is the maths?' to both their satisfaction and ours.

Junior school headteacher

The impact of the course, as I see it, was twofold. First, it provided a great deal of intellectual stimulation and engendered a new concept and approach to the teaching of mathematical skills. It made me question the relevance of the kind of mathematical problems we were setting our children.

Secondly, the impact upon the school was immediate and generated interest from the children, the rest of the staff, and the parents. This was

Mathematical topic	% of projects in which the topic occurred
Counting	90.5
Addition	90.5
Subtraction	81.0
Multiplication	76.0
Division	28.5
Fractions	14.3
Ratios	9.5
Percentages	14.3
Measuring (length, area, time, weight, etc.)	95.0
Comparing	71.5
Estimating	81.0
Organizing data (e.g., making a tally, ordering)	90.5
Surveys	71.5
Statistical work	38.0
Making graphs or charts	76.0
Spatial awareness/representation	71.5

achieved because everybody concerned had a vested interest and could see the results within the physical limits of the school site. We all had a will to succeed and because of the personal involvement it was possible to see mistakes and find alternative solutions. The problem-solving crossed much of the curriculum; old skills were used and new ones acquired and put into practice. The children could see the necessity for the acquisition of new skills; they had a problem which they could not solve because they did not know how to proceed. They learned with more understanding and certainly with more application, so that they could proceed with solving their problem.

Children learned to listen to each other, take notes, report back, evaluate situations, write letters which necessitated and received replies, and ask pertinent questions. They learned to look for alternatives and to find out at first hand that problems can have many and varied solutions. Their mathematical skills increased; it was most important that measurements and calculations were accurate and the children could see why. Their estimations of length and costing were important and meaningful. Pictorial representation was used with effect in portraying data. The children could see the purpose in observing closely, noting what they had seen, and discussing the implications.

We are all proud of the end result and we learned from our experience. The children learned much that was relevant to them in their everyday lives. The simple fact that they were able to talk through problems and exercise social skills was a big step forward. They used their abilities to help each other and, since writing was not always essential, everybody was able to contribute in a realistic and positive way to the project. The gardening skills were tried and practised with great enjoyment and for many of our children this in itself was an unknown experience and joy. The responsibility to continue with the project has been willingly taken on by the whole school. We are now preparing to meet and tackle our next problem.

A direct result of the 'real problem-solving' project was that we examined the impact it had made upon the teaching of mathematics. It seems to me that the difference in this experience was essentially that children were acquiring and learning new skills because they could not proceed without them, whereas previously they were being taught new skills mainly when the teacher thought they needed them and were ready to learn. The children in the problem-solving situation could therefore see at first hand that it was essential to learn in order to proceed.

Mathematics became not merely a series of exercises to prove that the child had learned the new skill, or a practical experience that bore no relation to the real world. If the child had not learned he could not proceed; moreover he was using mathematics in a live situation. The children also had to use other skills, both well practised and newly acquired, alongside their mathematics to enable them to solve the problem. Mathematics was now demanding that other areas of the curriculum were used in close relationship with it, instead of being an incidental tool for other topic work.

Real problem-solving had not only an impact upon the content of the curriculum but upon the demands both children and staff made, and the whole approach to learning and our attitudes towards it took a different perspective. [X]

The Mathematical Association's Diploma in Mathematical Education

Traditionally in Britain, in-service qualifications for teachers have been offered by universities and colleges, rather than by professional associations. The Mathematical Association has, however, broken new ground by initiating national qualifications in mathematics at diploma level for serving teachers. A part-time diploma for underqualified secondary mathematics teachers was popular in the 1960s, but lost ground when the Open University and other institutions started to provide a wide range of mathematics courses for part-time students. Recently, this Mathematical Association diploma has been replaced by a diploma designed to fill another national need – that for upgrading the mathematical qualifications of primary school teachers.

The contributors to this section are the chairman of the diploma board of the Mathematical Association, a lecturer who teaches on one of the courses, and a student.

The chairman of the diploma board

The year 1978 was an important one for British in-service education in mathematics, for it witnessed the launching of a major new qualification, the Mathematical Association's Diploma in Mathematical Education. The association's previous diploma had been designed as a mathematics course for secondary teachers with nationally set examination papers. The new diploma was to be a judicious blend of mathematics and education, designed to equip suitable and interested teachers of five–thirteen-year-olds to make a strong personal contribution to the teaching of mathematics within their own schools. It was significantly different from the previous course in other respects too, mainly because the association had opted for a validation model of assessment, not unlike that used by the CNAA in its validation of degree courses in nonuniversity institutions. This paper outlines the organizational features of the new diploma, which is administered nationally through a diploma board of fifteen members drawn from a wide range of geographical, institutional, and other backgrounds. For some years there has been concern about the quality of the mathematical knowledge of many primary school teachers. The recent changes which have affected classroom teaching – relating to new topics and practical approaches – have left many teachers confused. Also the rapid expansion of the teaching force in the 1960s – to cope with a bulge in the birth-rate – has meant that the mathematical standards of new recruits were often very modest.

Thus, the new diploma aims at improving the mathematical knowledge of teachers. Of the 200-hour contact time stipulated as a minimum requirement, approximately half must be spent on mathematics, and this is assessed mainly by written examination. But the intention throughout is that the mathematics taught should be linked to the teaching in the classroom; although the assessment of the course may be separated into convenient headings, the teaching should be seen as a unity.

Diploma courses are situated in individual colleges throughout the country, and each college develops its own course to suit its own circumstances and teaching strengths.

The diploma board has issued several sets of guidelines. For example it produced (initially) a model syllabus and its assessment guidelines insist that successful candidates must satisfy the examiners in four aspects: mathematics, a mathematical investigation, mathematical education, and a child study. Based on these guidelines and on personal contact with a member of the board, colleges submit a proposal which includes the teaching programme, the assessment programme, and details of staffing and other resources. Members of the board visit the college and, when satisfied, approve the college submission and an external moderator nominated by the college. Hence each college has its own distinctive package within a general framework.

A national standard is achieved through regular meetings of moderators and lecturers. By this means, the board transmits its philosophy on assessment and ethos. The ethos is most important: the diploma must be seen as an appropriate second-level course suitable for experienced teachers. A course is of little value if it does not relate to the teachers' needs and maturity.

No college submission is approved unless there is a genuine commitment to integrating all aspects of the course: it is not acceptable to combine a number of disparate units, nor (for example) to teach all the mathematics first. Generally courses are part-time, spread over a two-year period, throughout which the board expects colleges to maintain a proper balance of work. The implementation of the programme is checked through regular contact with the external moderator and board members.

How has the diploma been received? Certainly it has been successful in recruiting teachers. In the first year (1978) over six hundred teachers started courses at over forty colleges. In the second year (1979) a further six hundred started courses and the number of colleges increased to fifty. One of the aims is to provide trained leaders so that primary schools have the opportunity to improve standards from their own staff. The board is pleased to note that many of the first recruits have had more than the minimum two years' teaching experience. Indeed, many heads and deputy heads have enrolled as students studying for the Diploma in Mathematical Education. The first two years of the diploma have been most encouraging. [Y]

* * *

A college of education lecturer

The Mathematical Assocation's Diploma in Mathematical Education was a new type of course for my college: it was to be a cooperative venture staffed by a team of tutors in mathematics and mathematical education from three institutions of higher education in the locality. We planned our parts of the course: 'Mathematics', 'Children and their Mathematics', 'Applications of Mathematics', and 'Investigations'. As well as these units, which were geared to helping the mathematical needs of teachers of the five–thirteen age-range, teachers would make a child-study or classroom-study, focusing in detail on some aspect of their children's work. The course would take place over a two-year period, three hours on one evening a week, together with four full-time courses of two days' duration during school holidays. The teachers would be expected to read and to write an essay each term, as well as working at their mathematics, their child-study, and their investigations. We decided not to have a single final mathematics examination, but a short examination at the end of each year, so as not to place too great a burden of examination revision on the teachers.

Having planned the course, we wrote an explicit and detailed brochure, printed posters and application forms, and sent them round to every primary and middle school within a thirty-mile radius, and sat back to await the flood of applications. Everyone said this was the course primary teachers had been waiting for. However, there was no flood of applications. There were applications from thirty-five teachers. We interviewed them and offered places to seventeen teachers, the maximum number we could manage. They are exceptional primary teachers, with a high level of commitment to mathematics. They are able, articulate, and enthusiastic. They teach from the reception class to the middle school. Several are graduates, and their backgrounds are in a wide variety of subjects. One gave up mathematics at thirteen, some took mathematics at A-Level or as a main subject at college. One has a science degree. Working with them is demanding, stimulating, and enormous fun.

The first course is now halfway through. The original seventeen were reduced to thirteen in the first term, eleven women and two men. Sadly, we lost the two headteachers of small schools, who found the pressure of a regular commitment of this magnitude too much for them. The workload has become clear – opinion among course members is divided as to whether it is quite as heavy as an OU degree. Essays are carefully prepared, and many of the course members have read widely in mathematical education. The mathematical side of the course has been varyingly received – it was overambitious for some, while others could have progressed faster and covered much more ground. Certainly the first-year mathematics examination caused anxiety among a considerable number of members of the course. The tutors were much more anxious about how teachers would take to doing their own mathematical investigations, but here we had reckoned without the willingness of primary teachers to 'have a go'. No sooner was the first problem propounded for group investigation – 'If you make a plane cut through a cube what shape might the cut face be?' – than each teacher turned to her neighbour, seized a cube, and began to work. We have not yet asked the group to do individual mathematical investigations for assessment, but their willingness to look for patterns and follow up their own ideas is most encouraging.

And what do the group members think of the course? I asked one of them to write down her views. Here is her evaluation of the two components of the course.

1   *Children and their mathematics*
    This aspect of the course is proving to be of great interest and practical

value. All the topics covered to date have been entirely relevant to the ordinary classroom situation and to the problems of individual children. The work has been of value to both infant and junior teachers and the link between junior and secondary schools has been considered. This aspect of the course would be of great value and interest to any teacher concerned with the teaching of mathematics; it has been both thought-provoking and stimulating.

2 *Mathematics*

From my own point of view this component of the course has been far from satisfactory. The members of the course are taught as one group and therefore each lecturer has to assume that the members have already reached a certain standard of proficiency in the topic being worked on. This is not so: some course members studied mathematics as their main subject at college, others have A- and O-Levels and one or two, myself included, have no qualifications at all. Consequently, being one of the members with no qualifications, I find myself a nonstarter. We are encouraged to ask questions, but being way behind those with A-Levels a few questions in the middle of the session do not really solve my problem – it may also hold back the more mathematically able members and make the sessions tedious for them.

It seems strange to me that at one moment we are hearing of the work of Piaget, Bruner, and Bell, who speak of people being at different mental stages and of the inadvisability of assuming knowledge in a pupil, and the next moment we are being taught as one large group with the assumption that we already have a certain level of knowledge, in preparation for the same examination.

If this aspect of the course is to be meaningful for all members, then either there should be a minimum qualification for entry in the first place, or work should be prepared on an individual basis and members should be able to sit examinations appropriate to their own ability and not one general examination. [Z]

Other course provision

It would be impossible to cover the variety of other in-service courses provided by institutions of higher education, and by voluntary bodies and loose associations of teachers. This section provides two examples. The first is a voluntary group of teachers, who originally came together through the initiative of the Nuffield Mathematics Project, but who have continued to work together and to spread their ideas. The second example shows how a college can work as a catalyst in a new area, in this case computing, by providing know-how and support, and how the college staff also gain from the work.

A curriculum development leader

At the beginning of this decade the Nuffield Mathematics Project was coming to the end of its existence, and attempts were made to set up regional bodies to continue the spread of good practice in primary mathematics. It was realized that change in education does take a long period of time to evolve. The Nuffield ideas had only just been born, and considerable support was needed in the regions to nurse the baby through infancy into childhood and beyond. The Northern Mathematics Council (NORMAC) was the outcome of this exercise in Northwest England (which has a population of about nine million). Since its inception NORMAC has grown from strength to strength. Sixteen local authorities and ten colleges of higher education are affiliated, and personal membership is also available. Two residential courses are organized each year and the response from teachers has now far outstripped the available places. By using several hotels in a seaside resort out of season for residential accommodation, and meeting in a large thirty-class primary school, NORMAC now accommodates up to 180 people on each of the annual weekend courses. At first two courses (one infant and one junior) were arranged, and now four courses are on offer.

A recent programme indicates the type of course provided:

Course A – for teachers of children in the age-range five–eleven years
Number from pattern and activity:
The course will deal with number work and number patterns arising from practical activities. Starting points will include aids such as logic blocks, games, and electronic calculators.

Course B – for teachers of children in the age-range five–seven years
Shape in the infant school:
The course will consider:
   1   the experiences of, and the reasoning behind, 'junk' modelling;

2   repeating patterns, tessellation, and tiling in daily life.

Teachers should bring four containers of different shapes, two of them with lids, and some examples of printed cloth or paper showing repeating patterns.

Course C – for teachers of children in the age-range seven–eleven years

Pictorial representation and graphs:

The course will consider relationships which are fundamental in mathematics. Relationships which can be expressed in words, symbols, and pictures, particularly those relating to the environment, will be investigated. The practical content of the course will be directed towards the mathematics children learn, the skills they acquire, and new avenues for development.

Course D – for teachers in the age-range seven–thirteen years

This course will be useful for headteachers, for teachers in charge of mathematics, and for teachers who in the future might assume this responsibility.

a   *The curriculum (three sessions)*
   Methods of balancing the curriculum. Planning the whole curriculum. The role of the specialist, particularly in view of the recent primary survey.[8]

b   *Assessment in mathematics (three sessions)*
   The function of assessment. Commercially published tests. Assessment as part of teaching.

The speakers are usually national figures who arrange their courses in six sessions. The social side of a residential course is also highly valuable, giving rise to a complex network of friendships and contacts, which has resulted in much interaction between different schools and local authorities.

Many teachers have to pay their own expenses for these courses (at present approximately £35) as some local authorities cannot provide finance for in-service education. The converse is also evident, as some local authorities send teachers as representatives, then use them to assist in their own in-service programmes.

NORMAC also subsidizes courses organized by LEAs. The finance for such ventures comes mainly from the sale of eight different pamphlets (twenty-five to forty pages) which have been written by NORMAC members for the assistance of the classroom teacher.

The Nuffield Mathematics Project was reborn in 1977 to produce pupil material, and fresh attempts were made by the National Committee (NATMAC) to revitalize other regions. This has proved successful and other MACs are now operating successfully in their quest to offer some support for the classroom teacher. [AA]

*   *   *

A college of education lecturer

There is a great diversity in the type and degree of commitment to computer education among local education authorities in England and Wales. The college of higher education in which I work lies within thirty minutes' travelling time of half a dozen LEA areas and is involved in in-service work in computing with teachers from this region. The experience of the college in this field is an example, rather than a sample, of the activity that is taking place in the country. In Birmingham and inner-London, for example, far greater resources have been devoted to the stimulation of the growth of computer education than is the case in the area surrounding this college. Even within the area served by the college, there is great variation. One local LEA provides most secondary schools with some facilities, usually a micro-computer and a terminal to link to a large central computer, and a number of in-service courses are organized at a central further education college. At the other extreme, another LEA offers much less assistance, the facilities which schools acquire being mostly through the exertions of parent-teacher organizations.

At the present time, the number of teachers with knowledge and training in computing is very small, and the rate of entry into the profession of such teachers is very low, so that the growth of computer education in schools, which has been swift (about 10% of secondary school children are exposed to computers today against almost none ten years ago), has been almost entirely maintained by in-service training of teachers, and this will continue to be the case well into the future. In addition, the rate of technological change has been such that those teachers who already have some training need constantly to update their knowledge.

It is within this environment that the college attempts to:

1   provide teachers with the skills necessary to make the best use of the facilities they have through in-service courses;
2   supplement the meagre facilities available to some schools;

3   encourage schools which at present have no involvement in computer
    education to introduce the use of computers as part of their cur-
    riculum by direct involvement of college staff in the schools;
4   provide a forum for discussion for teachers.

The in-service courses given by the college involving computer-based
work are of three types:

1   A one-year part-time Certificate in Computer Education course is
    offered, involving about one hundred hours contact time plus further
    time for individual project work. The course aims to enable teachers
    to become sufficiently competent programmers to be able to write
    simple teaching programs and packages for themselves, to be able to
    evaluate packages from other sources, to have a reasonably wide view
    of current applications of computers and some understanding of
    relevant technological details, and to appreciate the implications of
    this for society as a whole and education in particular. Intending
    specialist teachers in the sixteen–nineteen age-group are expected to
    undertake more advanced study elsewhere. The teachers attending
    the course are almost exclusively young mathematicians. This is not
    by any policy of the college; indeed we had hoped that groups would
    have been to some extent multidisciplinary and that teachers with
    appropriate skills from subjects where there is a surplus would be
    attracted to computer education. This has not proved to be the case,
    and it has therefore become our aim to help schools create a pool of
    expertise within their mathematics departments, whose teachers can
    then give encouragement and assistance to other departments when
    their interest is aroused.
        The flow of knowledge in this course has by no means been a
    one-way process, and the material produced by course members has
    been of value to the college, and the ideas generated have been
    disseminated to other schools through the college.
2   Short courses of ten–twelve hours' contact time are designed to satisfy
    particular needs suggested by schools or identified by the college.
    They have not proved to be of wide appeal but have sometimes
    allowed specific work to be done with a group of teachers from a
    particular school. For example, the college uses micro-computers for
    a number of administrative tasks and a recent course, 'Using
    Mathematics for School Administration', has led to help being given
    to a school which wishes to justify the purchase of a micro-computer

by using it for administrative as well as educational purposes.

3   Many of the in-service mathematics courses such as the Diploma in Advanced Studies in Education (Mathematics) course and the Primary Mathematics Certificate include a small module of computing, providing a short introduction to programming and to the possibilities opened up by the use of a computer in schools. A number of primary school teachers have become enthusiastic and are using, or planning to use, a micro-computer borrowed from the college in their classrooms.

All teachers who attend in-service courses involving some computer work, and who wish to introduce or expand courses in their own schools, are offered assistance with the provision of facilities and teaching materials. Micro-computers or a portable terminal are provided on loan for periods of between a week and a month at a time, perhaps once a term, and two telephone lines to a mini-computer within the college are reserved for schools' use. The provision of these facilities is maintained after the completion of courses and thus fruitful contact and dialogue is maintained, effective feedback on the success of courses is gained, and ideas for future courses and meetings generated. Such meetings are usually organized under the aegis of a local branch of the Computer Education Group which meets in the college.

Some years ago the college noted that a number of schools wished to introduce computing courses but their staff felt that they lacked the expertise necessary to take such a step without support. Accordingly, each year a member of staff from the college has undertaken to spend approximately half a day per week teaching in such a school. Staff from the school concerned, who would usually be taking part in a course at the college or elsewhere, have observed and cooperated in the presentation of lessons, and eventually the school has taken over responsibility for all the teaching. This exercise has been mutually beneficial. The introduction of new courses in the schools has been much smoother than it otherwise might have been, while the college lecturers have maintained first-hand contacts with the classroom; this has contributed greatly to the success of their college-based in-service work. A good relationship has also been forged between the schools and the college. Obviously this type of assistance can only take place in the case of a very few schools but it is nevertheless a very important aspect of our work.

In conclusion I would say that, despite its limited resources, the college has been able to make a very positive contribution to the development of computer education in schools in its area through the in-service work described. We have come to the view that the close involvement of lecturers with work in schools and the provision of adequate facilities to enable schools to initiate courses is an essential part of any successful programme. [BB]

CHAPTER 8

THE VARIETY OF PROFESSIONAL
DEVELOPMENT

In previous chapters, we have shown how teachers of mathematics can develop professionally by working together with their colleagues in their school, or by attendance at in-service courses which may be provided in the LEA or by a variety of other agencies. There are other activities by means of which a teacher of mathematics may engage in professional development in a more informal way. Taking part in the activities of a professional association, reading journals, and taking part in the setting and administration of the public examinations which are set for pupils are three ways of doing this, about which teachers have written in this chapter. Finally, a teacher who has taken part in liaison with the industry in which some of his pupils will work, and who has studied the mathematics of the shop floor – or perhaps in this case the kitchen floor – tells us some of the needs for mathematics which he has found that employers actually have. No doubt his lessons on percentage will continue to be enlivened by the necessity that Smarties should not be packed in damp tubes.

Teachers as examiners

The majority of children in England and Wales take one of two public examinations at the age of sixteen; these examinations are the O-Level examination of the General Certificate of Education (GCE), taken roughly by the most able 20%, and the Certificate of Secondary Education (CSE), taken by pupils down to the 60th percentile or lower. The involvement of teachers in the GCE is not great; although some teachers work for additional payment during the summer by marking GCE scripts which pupils have written from the examination papers set by a very small number of examiners.

The CSE, however, is a teacher-controlled examination and, because it involves large numbers of teachers in its organization, it has proved a potent force for professional development for secondary school teachers. It gives an opportunity for professional discussion of syllabuses, standards, and problems between teachers from different schools, and it provides teachers with the possibility of examining the work of pupils in schools other than their own, comparing the strengths and weaknesses of their own work with that of colleagues elsewhere. The CSE is organized by regional examining boards, so that comparatively large numbers of teachers take part in its administration and examining.

The CSE examination can be taken in one of three modes, according to the school's choice; the most popular of these are Mode 1 and Mode 3. Mode 1 is an external examination; pupils follow a prescribed syllabus and take an examination set by an external examiner – this model is the same as that usually found in the GCE. By contrast, Mode 3 is an internal examination, externally moderated; a school devises its own syllabus, which has to be agreed by the board, and sets its own examinations (or tasks for continuous assessment); the assessment is moderated by the board, with the object of ensuring that a pupil who receives Grade 1, or Grade 4, is in some way comparable with a pupil who was examined by another mode, in another part of the country, and following another syllabus.

The account which follows is of a meeting of the mathematics panel of a regional CSE board.

An institute of education tutor

The meeting took place in early October at the teachers' centre in a small town in the East of England. It was typical of many such meetings being held up and down the country. The twenty or so teachers present had travelled up to one hundred kilometres. They came as elected representatives of groups of teachers in their own areas, and their schools had allowed them a day's leave of absence so that they could attend. There was just one administrator in the room.

The meeting began by reviewing the examination just completed the previous summer. It was noted that although the Mode 3 syllabus at —— High School was satisfactory for pupils of lower ability, the examination papers were not of a high standard; the representative for the area agreed to visit the school and to encourage the teachers there to improve the examination. It was remarked that at —— Village College an unusually high proportion of candidates were awarded the highest grade; but the teacher responsible for moderating the Mode 3 examination in this school was able to assure the meeting that the students were of very high ability, and that the grades were justified by the work presented.

It was not long before the meeting turned its attention to the 'burning question' of the day – had the possession and use of pocket calculators given some candidates an advantage in their examination? Some not very conclusive statistics suggested that it had made little difference in the test of low-level skills, but that students with calculators had done better in problems; but then, it was argued, it was probably the better and more highly motivated students (the ones who would do best in problems) who bought their own calculators. Perhaps, in any case, this was only a temporary difficulty; maybe in two years' time all the students would have their own calculators, or they would be provided by the schools. The meeting decided that this was a problem to which there was no simple solution, but agreed to keep it under review.

The morning was well advanced before the meeting came to the main business of the day – to plan next year's examinations. First, the committees had to be chosen to 'moderate' the central Mode 1 examination – that is, to check the correctness, suitability, standard, and presentation of the question papers. A balance had to be struck between the wish of all the teachers at the meeting to be involved, and the need to have a small committee to conduct the business efficiently. Then came the consideration of the thirty or so schools which had elected to set their own Mode 3 examinations, based on their own syllabuses and often involving assessment of work done by the pupils throughout the course; for each of these schools individual moderators had to be appointed, from among the teachers present at the meeting, who would ensure that the standards of these examinations were comparable with those of the central examination. During the coming winter this would involve many hours of working through questions, discussions with teachers in the school, reading and commenting on pupils' 'projects' and examination answers – all in addition to the teachers' regular commitments in their own schools.

Finally, there were new syllabus proposals to be scrutinized. This school's proposal needed more precise description of the assessment of pupils' course work; that school wished to modify its syllabus by introducing an extra optional topic. Each proposal was discussed in detail, not just as the basis of an examination but in its presentation and in its viability as a teaching programme, and was accepted only when the meeting was convinced of its quality.

The CSE is an examination with a difference. It has its roots within the schools, and is administered on democratic principles. Teachers are encouraged to develop their own syllabuses, but are required and helped to produce them to agreed professional standards. To this end, they have to engage in dialogue with other teachers from outside their schools about the details of their teaching and assessment programmes.

The result has been, in the fifteen years of the examination's existence, a remarkable change in the attitudes, responsibilities, and experiences of many secondary school teachers. No longer are they confined within their own classrooms; motivated by the interests of their pupils, they have accepted a cooperative role in syllabus development and an ultimate accountability to their colleagues in the profession at large and, more widely still, to the society which the examination serves. In this way, the examination system has had an important part to play in the professional development of teachers. [CC]

Professional associations

It is one of the hallmarks of professionals in any area that they band together into a professional association. Often these associations are the guardians of professional standards and expertise, and are regulators of admission to the profession, and of dismissal from it. Sometimes they also bargain for their members' salaries or regulate their fees. In the teaching profession in England and Wales these functions are divided between different bodies: the teachers' unions, such as the National Union of Teachers, conduct salary negotiations; the DES grants qualified teacher status to those who satisfactorily complete their initial training; and subject expertise resides in a variety of associations of subject teachers. In fact, only a minority of teachers belong to these subject associations, but for those who do belong, they are a source of stimulation and a means of making professional contacts with like-minded teachers. They also provide journals which help to keep teachers informed about developments in the teaching of their subject. Through their national councils, the subject associations contribute to the development of educational thinking, and at local level their meetings provide a forum for personal interchange and work.

Of the two contributors to this section, one is a long-standing member of the council of one of the subject associations in mathematics, while the other edits a professional journal.

An officer of a professional association

Associations of teachers are a familiar part of the mathematical scene in many countries. In the UK we have two major associations of this kind – the Mathematical Association (MA) and the Association of Teachers of Mathematics (ATM) – and several smaller 'special interest' associations dealing with applications of mathematics, teacher training, computer education, etc.

It is interesting to reflect that neither of these two associations started as an agency providing in-service support for teachers. The MA was founded in the 1870s as a 'pressure group' to get school mathematics released from the stranglehold of university domination – in particular, of the required formal study of Euclid. The ATM began in the 1950s as a group of enthusiasts who met together to devise and promote the use of visual aids in teaching mathematics. But both have developed into large organizations of several thousand members with interests spanning the range of mathematics teaching from primary school to university, and providing an important resource for the professional development of teachers of mathematics. Indeed, for many years before World War II, when specialist teacher training was in its infancy and when most education authorities saw their responsibilities solely in terms of the provision of buildings and the employment of teachers, self-help through the MA was almost the only professional resource available to mathematics teachers in the UK.

The main activities of the association have changed surprisingly little over the years, though some activities have developed a different emphasis. Conferences have become less sedentary, and encourage more member participation in small groups and concentration on particular issues. Journals have to resist pressures to become repositories of recondite but irrelevant research, and to keep firmly in view their primary function of being useful to, and enabling communication between, teachers in schools. More local branches encourage teachers to bring their older students to meetings with them, an experience which often carries over into classes during the following week. New printing facilities have encouraged the production of small, informal publications on particular topics. In all these respects, however, the associations have continued to fulfil an important supporting role for some mathematics teachers.

The word 'some' in this last sentence is crucial, and emphasizes both the strength and the weaknesses of the associations as vehicles of in-service training. To become a member of an association remains a voluntary decision, and represents a commitment to mathematics teaching; it is estimated that only about one in five of all secondary full-time teachers of mathematics in Britain belong to either of the two associations. For these teachers, an association offers a sense of belonging. The journal is 'their' journal (though it is certainly read by many nonmembers in libraries and teachers' centres); the association speaks for 'them' (even when they have taken no part in formulating policy). Even with their present-day enlarged membership, associations continue to meet a personal need.

An example may help to emphasize this point. It was suggested recently that the government could influence mathematics teaching across the country at very little cost by buying copies of the journals of the two associations for all secondary schools. What was not appreciated was that such a step could well deprive the journals of the personal involvement of members on which they thrive. It is central to the role of the associations that the concept of membership should remain a reality; their strength lies in the commit-ment of their members, not in their numerical size.

In this lies the answer to the question of whether there remains a place for the voluntary association in an age when education authorities have accepted a responsibility for the continuing training of the teachers they employ. Some of the activities previously assumed by the MA and ATM – local meetings of teachers, for example – have declined as a result. But a takeover of their functions on too large a scale could deprive the country of one of its most important sources of leadership and initiative in mathematical education.

It would be wrong to imagine that the associations are in any real sense a means of making mathematical education more 'democratic'. In their early history, as small groups of teachers meeting together for a common purpose, it was possible to identify an 'association point of view'. It is ironical that nowadays, when the help of the associations is increasingly being sought – by bodies such as the government's Assessment of Performance

Unit, the Schools Council, or the Schools Broadcasting Council – their membership is too large and diverse for it to be possible to identify a consensus opinion of an association as a whole. Inevitably, in making a response to such approaches, only a few leading members are involved. Nevertheless, by debating national issues in national conferences, in meetings of local branches, or in publications, the associations do play a part in bringing major issues of policy to the attention of their members and offering opportunities, to those who wish, of developing their own ideas and making a personal contribution to the debate.

One problem which faces both associations in the UK is how to develop their role vis-à-vis mathematics in primary schools. About 80% of all teachers of mathematics in the UK work with children under the age of eleven, but few of these are in any sense 'specialists' and almost all of them teach across the whole school curriculum. It is rare for a primary school teacher to accept a commitment to one particular subject to the extent of wishing to join an association such as the MA or ATM; and yet the work which primary teachers do is fundamental, and they are an essential part of the 'body mathematical'. Both associations have sought to make provision for them through special publications, journal articles, occasional regional conferences, and the like; but it is participation, not provision, which is the justification for the existence of a professional association – and this is a problem which remains to be solved. [DD]

*   *   *

The editor of a journal

I think it was Dr. Johnson who said something about there being nothing like imminent execution for focusing the mind. I feel the same about writing. This is less drastic than execution, although to some teachers it seems to be almost as terrifying.

Even a local duplicated journal published by a mathematics centre (hopefully called *Feedback*) seems to instil fear. I am always on the lookout for articles, and when I meet a suitable idea I tend automatically to say, 'That's nice. Write it up for *Feedback*.' The response is usually, 'Who – me?' If the teacher is on a course at the centre I can put paper and pen in front of them and say, 'Yes – now!' But the fear of writing is a handicap anyway, and manifests itselfs most often in that archaic style encouraged by essays written at college, where any living ideas are stifled by educational jargon, stereotyped phrases, and irrelevant platitudes.

TOM—I

An interesting exercise occurs during the annual six-week full-time courses for primary teachers that we have run for five years. Once a week they visit a local primary school, the first week to observe and make arrangements, and in subsequent weeks to spend an hour working with just two or perhaps three children each. I ask them not to teach, but to present an appropriate situation, problem, or game and then to observe.

After the first such session I present each teacher with a spirit master and ask them to write what happened. There is consternation on their faces, but no one ever objects. The recent experience gives them the content of their writing, but the spirit master indicates that everyone will have a copy of each write-up and this gives them their audience, and hence their style, which is always informal and reasonably informative.

However, their first write-up has gaps, and – this is the point of the exercise – they realize they were not really attentive enough during that first session. The point is thus forcibly made, and the following week they return to their two or three children more conscious of what they are seeing and hearing, and incidentally more careful about recording or remembering the details.

The object of writing, then, is to clarify one's thoughts about what has happened. There can be far more to it than this, because firstly one must distinguish what actually happened from what one thought happened. Then one can consider possible explanations for what took place, different interpretations of actions and remarks, and possibly fit things into one or another psychological framework. The value of the immediate sharing of the accounts of experiences with others is that a discussion can then take place that helps the writer to clarify further all these ideas against a background of other people's justifiable misunderstandings, misinterpretations, and alternative hypotheses.

The question of publication of such material in this context is almost irrelevant, except that imminent or eventual publication pinpoints an audience to whom one is trying to justify one's ideas. This affects the *style* of writing, as it seemed to with the teachers described already (although I wish more authors of articles intended for national publication bore in mind the type of readership involved!). But it also affects the *level* of discussion.

We are conscious of both these points far more when we use spoken words, and we vary our speech according to whether we are talking to younger children, older children, parents, primary teachers, secondary mathematics teachers, headteachers, employers, or people involved in pre-service or in-service education of teachers. We seem to be less conscious of them when we are writing.

By the way, what I am saying also applies to children's communication, but the differences between subjects are marked here. English teachers seem to be aware that any piece of writing should be directed at someone or something in particular – fellow pupils, the teacher, one's own parents, parents in general, the school magazine, the wall newspaper, the local newspaper, a national newspaper, a prospective employer. If only we provided an actual audience for children's written mathematics, it might become more coherent. And in particular we should distinguish between that recording of mathematics that communicates with someone else, and that which enables one's own thoughts to be sorted out.

The disadvantages, for teacher-writers, of a national or international journal are firstly that the audience is less well defined, and secondly that the feedback from that audience is at best delayed and at worst unknown. Here it is the responsibility of the editor to act as an intermediary on behalf of the readership, and it is here that I find myself often indulging in the same in-service education/professional development exercise as my groups of teachers, in which both I and the author learn from each other in a private dialogue that may go on for a considerable time, and which may result in extensive alterations to the article before eventual publication, or perhaps in no publication, either because agreement could not be reached, or because the dialogue itself was enough to satisfy the needs of the author in clarifying his thoughts. (Occasionally the result is publication in another journal!)

Indeed, the motivations of authors to put pen to paper are many and various. There are those who – in the words of my college lecturer in English years ago, G. Orton-Smith – 'have to say something' rather than 'have something to say'. He was referring to the external compulsions on children to write, and we could consider his thoughts in relation to children's mathematics too. But teachers and educators generally have both internal and external compulsions. The author bursting with ideas wants to get them down on paper as though he would literally burst if he did not; the editor must decide between his need to write and the readers' need to read, and there are in fact times when the balance is difficult, and it is important to publish an article for the sake of the writer rather than the reader – other things being almost equal! And, let us face it, there are those with external pressures to publish, though this is fortunately not a characteristic of educational institutions in the UK; the editor can usually recognize the results of this pressure (often typified by three duplicated copies – as though there were many others in circulation – and sometimes even a duplicated letter with the name and address written in by hand) and must still consider each article on its own merits, even though it may be with less sympathy!

There are particular motivations, like publicity for a particular project, either with the best intentions of sharing some ideas, or with the worst ones of selling some materials. There are axes to grind, flags to wave, bandwagons already jumped on, or offers for others to be taken for a ride. Although there are always more readers than writers, and the editor must protect the former, his role as an agent for development must also apply to the writers, distinguishing their professional considerations from personal ones, but if necessary acting to influence both in order to affect the professional ones.

There are times when writers like to talk to each other, and the journal is a forum for a debate, held in full view of the entire readership. For example, a long-term debate was conducted in *Mathematics Teaching* on different types of understanding since R. R. Skemp's article in December 1976, and a shorter one appeared in June 1977 as a result of T. Leddy's article about the teaching of integers in the previous issue. The level of discussion tends to be higher (whatever that means), and the editor's task is to prevent it from becoming esoteric. The debates are obviously intended to be public, otherwise the authors would merely write letters to each other – though there are times when the editor arranges that, if it is more appropriate.

Ideally, those readers/members who are regular writers share the same aims as the editor has in producing the journal. There are comparatively minor aims, which should perhaps be called objectives, but I am trying to avoid here the same educational jargon that I delete from articles. A journal should inform about current ideas, trends, practices, projects, materials, and books, giving opinions about these, and giving opinions anyway Perhaps the idea is to avoid being *un*biased. Views are important, and are put to be considered rather than accepted. It is axiomatic that the readership is educated!

Somewhere in this list feature practical ideas about teaching, 'teaching tips' if you like. I have an uneasy feeling about these, as though there were something aesthetically wrong about a section, say, on ideas for lessons, suggested workcards, ten things to do with a paper clip, the 'best' way to teach subtraction. Somehow such obvious spoon-feeding does not fit my main aim, reveals a paucity on the part of authors who cannot or will not say more about what happened when they tried an idea on their own children, and insults those readers who look for something more stimulating. At the same time, I remember in my early days of lecturing to teachers the advice given by Cyril Hope, 'Offer them all the philosophy you like, but give them something to teach on Monday morning!'

My main aim when working with teachers is to stimulate them to think about their teaching. I may, indeed, offer them 'tips for teachers', but in such a way that they are stimuli to thought. I may question what they are doing, ask them to justify it, suggest alternatives. What one cannot do is to provide an alternative curriculum without a commitment to a philosophy behind that curriculum, and by curriculum I mean methods and attitudes and aims as well as content. So, the only way to effect curriculum development is to affect the teacher.

Given a philosophy of in-service education, then editing a journal fits naturally into it, as an aspect of in-service education. The personal nature of the philosophy may give a very personal bias to the journal, and it is not for me to say whether or not that is successful. That it is possible is due to the absence of editorial committees and panels of referees, and an authority invested in the individual by an association which chooses to organize affairs on the basis of responsibility and trust.

Any association allows for a whole spectrum of degrees of commitment. The commitment of an editor is great, but no less perhaps than the commitment of some of those who write. The least is a commitment to read, and that too is important. [EE]

Links with industry

In the last few years, there has been frequent criticism of the fact that school mathematics does not prepare young people for the mathematics they will use at work. Some of this criticism has certainly been based on the ignorance of both schools and industry of what each other does, and on the ignorance of both teachers and managers of what mathematics is actually used on the shop floor. However, this has led to great growth of informal local school-industry working groups, and these groups have become a source of professional development for many mathematics teachers.

Smartie maths, by a secondary school teacher

In the past few years local liaison groups representing teachers and employers have become quite fashionable in the UK. Some seventy such groups are listed by Professor Bailey.[12] What follows are the impressions which one teacher belonging to one of these groups has gained from touring one factory.

An early project of the York working party on 'Maths in School and After' involved the teachers in trying to identify the mathematical skills actually being used by new employees, mainly sixteen-year-olds, in local firms. In my case the firm visited was a chocolate manufacturer which is one of the biggest employers in the area. In all I visited six departments chosen as likely to require mathematical skills: central calculating, drawing office, planning, operational research, work study, and quality control. The computer department was not included.

What qualities do senior members of staff look for in the new entrant? 'An inquiring mind', 'ability to communicate', 'reliability, speed, and accuracy', and 'ability to develop a feeling for the correct size of quantities and to detect nonsensical figures' were mentioned more than once. This last skill, concerned with sensible approximation and estimation, was one which colleagues found was stressed during their visits to other businesses.

Another type of skill frequently required is that of extracting information from graphs, plans, tables, and scales. The reading of scales on a micrometer or weighing machine was an obvious example but in the work study department a nomograph was in use to give the optimum number of spot checks, while the times at which these checks are carried out are determined from a table of random numbers.

The fact that much information is presented in graphical or tabular form probably explains why algebra was not often in evidence on our factory tours. Where algebra did appear explicitly it was usually as a formula in which numerical values are to be inserted. The drawing office at the chocolate factory also cited the transposition of a given formula as an important skill. Some examples came to light of algebraic formulae containing 'awkward' features: capital and small letters in the same formula, a variable represented by two or three initials or even by a numerical symbol, the use of subscripts – the sort of things a mathematics teacher is likely to avoid at an elementary level for the sake of clarity and simplicity.

Metrication is still a bone of contention between some industrialists and teachers in this country and, judging by government reluctance to press forward, it will remain a source of confusion for many years yet. Since the early 1970s most school texts and public examinations have exclusively employed metric units yet it is still likely that a school-leaver will meet imperial units in his work as well as in everyday life. At this firm inches and millimetres, pounds and kilograms, drams and grams were all to be found in regular use.

And what of computation? How is the world outside the classroom doing its sums? The answer seems to depend on personal preference. Electronic calculators were in evidence in most offices but a slide rule was in use in quality control and at least one senior man still advocated pen and paper for long divisions. For me an interesting discovery was the comptometer in the central calculating department.

In this office a vast number of simple calculations, mainly additions, have to be carried out with speed and accuracy. Staff are trained to use a comptometer, which is an adding machine worked by a keyboard rather like an electric typewriter. It has two speed advantages over an electronic calculator: several digit keys can be pressed simultaneously and there is no '+' key to be pressed between successive entries. The machines can also be used for subtractions and multiplications, but these are more usually done on a calculator.

Under the heading of 'basic numeracy' it was percentage that made the most frequent appearance. In contrast to the convention taught in schools (*cost* price = 100%) it is the company's *selling* price which is reckoned as 100% in central calculating; a planning department calculation allowed 20% wastage when ordering stocks of cellophane-type material used in wrapping outer packages; work study considered the effect of a 4% weight increase on the job of manhandling cases of chocolate; and quality control have to check that the moisture content of the card used to make container tubes for Smarties remains under 8%.

These and many other examples of mathematics used by young employees in York eventually appeared in a pamphlet,[13] the tangible fruits of those brief but interesting contacts with the world of industry and business. It is more difficult to gauge to what extent the exercise influenced attitudes. Two years later the impression remains of the need to give our pupils experience of a wide variety of tables, scales, and graphs, to stress the importance of percentages and to hammer home the old instruction 'check that the answer makes sense'. But as for imperial units, here is one teacher who cannot bring himself to put that particular clock back. [FF]

CHAPTER 9

TRENDS AND PRIORITIES IN MATHEMATICS TEACHING

The changing role of teachers of mathematics

In the past, many teachers of mathematics in grammar schools entered teaching directly from a university mathematics course, with no professional training as teachers. This form of preservice education reflected the traditional idea that the task of a mathematics teacher was to pass on his own mathematical knowledge to the small proportion of the population who were selected at the age of eleven to receive an academic education. The mathematics teacher's tools for carrying out this task were the blackboard and the textbook, and he operated by demonstrating techniques, giving proofs, and setting his pupils exercises, whose methods of solution he usually demonstrated on the blackboard. His pupils entered the grammar schools equipped with arithmetical skills in the four rules of number, money, measures, fractions and decimals, and often with high hopes based on the fact that they had passed the '11+' examination. In fact, those mathematical hopes were only fulfilled in a proportion of these able pupils, as witness the many educated adults who 'never could do mathematics'. The grammar school teachers worked in an era of comparative educational stability, when the picture of mathematical education as the passing on of received knowledge and skills to willing pupils (in decreasing numbers as the age of the pupils increased) was an acceptable one. In the last quarter of a century, this picture of the task of a teacher of mathematics has become totally outdated.

No longer are teachers of mathematics found only in secondary schools; every primary school teacher is now a teacher of mathematics. Previously, primary teachers were expected only to be proficient in drilling their pupils in the skills of 'mental', 'mechanical', and 'problem' arithmetic. Now it is hoped that they will provide a secure mathematical foundation for all their pupils.

The changed expectation in the primary school is matched by a changed expectation of secondary mathematics. 'Secondary education for all' now means mathematical secondary education for all, up to the age of sixteen. The idea that mathematics is of such importance that all pupils should study it for a full eleven years, between the ages of five and sixteen, is of very recent date. Yet it is already true of the vast majority of pupils; the recent HMI Survey of secondary schools[14] states that:

> Mathematics is a regularly timetabled subject for all pupils during the first three years of secondary education. The most common practice in the fourth and fifth years is to provide a subject option scheme . . . but almost without exception mathematics continues to be compulsory.

The task of teachers of mathematics has become enormously broader and more complex but they are not, of course, alone in experiencing this growth of complexity; much of it is shared with all teachers in primary and secondary schools. Taylor, in a study of European teacher education,[15] has identified four other major aspects of the role of the teacher, as well as his responsibility in transmitting knowledge and skills and in upholding cognitive standards.

First, the teacher is a moral and political agent of society; this task is, of course, the subject of ideological battles from which mathematics teachers cannot stand entirely aloof. Secondly, the teacher is an innovator. Mathematics has an important contribution here, because of its increasingly key role in a technological society. Teachers of mathematics are in the forefront of the development of computer education, as well as taking part in curriculum innovation in their own discipline. 'Education for change' is another important, if little understood, aspect of the work of mathematics teachers as innovators. Thirdly, teachers now have a cooperative role. Mathematics teachers can no longer work in isolation, each behind his closed door, without regard to their colleagues or to the larger community. The present interest in mathematics across the curriculum, and the calls for school mathematics to strengthen its links with the mathematics of working life, are examples of the need for mathematics teachers to see their work as part of a cooperative enterprise. Fourthly, Taylor points out that the teacher is an agent of social and educational equality.

Education is often seen as a way of providing opportunities which all may take advantage of; mathematics is in a very exposed position in the equality debate, as a subject in which individual differences in attainment are particularly visible. All these developments in the idea of the teacher's role have caused teachers of mathematics to have a much broader concept of their professional life. Some see themselves first as teachers, and secondly as teachers of mathematics; with others it is the other way round. The contributions to this book have emphasized the professional life of teachers as teachers of mathematics but, in either case, professional life in all its variety demands very much more than the simple act of transmitting mathematical knowledge to pupils. One thing that is certain is that even the act of transmitting mathematical knowledge is very far from simple, and needs study and research, together with the application of research in the classroom setting, to make it more effective.

The phases of professional life

In this book, we have portrayed professional life from the time of entry to the profession, but it is difficult entirely to separate professional life from the preservice preparation for entry to teaching. Professional life has to build on the equipment of knowledge and skills with which a teacher enters the profession. Indeed, some of the perceived problems of mathematics teaching stem directly from the fact that the knowledge and skills with which teachers enter the profession are inadequate to the task they are immediately called upon to perform. For example, only a minority of primary teachers start their teaching careers with a sufficient background of mathematical knowledge and understanding, and with an attitude to mathematics which is positive enough to enable them to tackle with confidence the range of work in mathematics found in many primary schools today. Similarly, few secondary teachers enter teaching with enough sympathy and understanding of slow learners to enable them to devise suitable learning experiences for some of the groups of pupils whom they will meet. And this does not only apply to the slowest learners; pupils of average ability can present problems to an inexperienced teacher who has naturally drawn from his own education an 'academic' view of mathematics, so that he may well respond to less able pupils by providing a watered-down academic course.

Perhaps some of these deficiencies could be remedied in initial training; it seems more likely that only a continued programme of professional development over a period of years is likely to have any real effect on teachers whose own education left them with a narrow view of mathematics and negative attitudes to it, or on teachers whose own experience of the less able has been limited to classroom transactions on teaching practice, often in situations of some stress. This, the first phase of professional life, that of *induction* to the profession, is a time of adjustment to school life and a search for solutions to problems which the entrant might not have foreseen, or with which he has not yet learned how to cope.

In the writings by new teachers in Chapter 2, entry to teaching was seen to be a time of strain and self-testing. Several of these new teachers were not supported to any great degree by their more experienced colleagues, and indeed the arrangements for the induction of new teachers proposed in the James Report[7] and the White Paper *Education: A Framework for Expansion*[16] have come to very little. However, two pilot induction projects were undertaken in Liverpool and Northumberland, and the report on these schemes[17] drew attention to a number of ways in which new teachers were found to benefit from a structured programme of induction into teaching. It was found that probationers' teaching performance was improved, and those 'at risk' were more quickly identified and helped. The Liverpool report states:

> There is a clear consensus of opinion within the Liverpool Authority that the well-being and, more importantly to pupils, the performance of probationer teachers has been improved by the positive steps taken to assist them. The improvement in probationer teachers' well-being is attested by the decline in sickness and consequent absenteeism and by the reduced rate of wastage due to premature resignations or failed probationary years. The reported decrease in incidence of probationers having unresolved classroom difficulties cannot be fully explained by an increase in the quality of recruitment.

The pilot induction schemes also had some unexpected by-products, such as the significant professional development experienced by the teachers who acted as professional tutors to the probationary teachers. There was also increased liaison and confidence between schools, teachers' centres, and colleges of education; and finally, many schools were stimulated by the scheme to look afresh at their own policies.

After the induction phase of professional life, the need for *continuing education*, formal and informal, remains important. This has two main aspects: there is *staff development*, the purpose of which is to help teachers to perform their work more effectively, so that the quality of the education service may be improved; secondly there is *further professional study*, where the emphasis is on the individual teacher, his career, and his personal needs. In mathematics, the two are often closely related; for example, when a primary teacher is sent on a course for mathematics coordinators, the emphasis will be on the job of the coordinator, and on tactics for carrying it out, but as one of our primary teachers [H] says, 'theory is definitely needed (in order to answer some of the awkward questions mentioned)'. The theory supplies both the individual teacher's need for further professional study, and the education service's need for developing its staff.

A further aspect of continuing education is that which has come to be entitled *retraining*. The idea of retraining is at present largely associated with preparation to teach another subject or another age-range, and so as a means of dealing with teacher shortages or falling rolls. Some few teachers of other subjects are at present retraining as secondary mathematics teachers, but the concept of retraining is a new one, and many teachers of other subjects are now teaching mathematics in secondary schools with no more retraining than a few ideas picked up from their colleagues. For many of them, the need for retraining is great, to enable them to work with confidence in a new field.

Moreover, almost all of the married women who have returned to the profession after bringing up families have done so with no retraining whatsoever for the changes in the schools which they had left ten to fifteen years before. Nowhere was this more traumatic for them than in the field of mathematics, especially in primary schools.

It is a commonplace of thought about industrial development that the pace of change is now so fast that many workers will need to retrain several times in a lifetime. This rate of change will not leave school mathematics unaffected, and the concept of retraining for new ideas and emphases, both within mathematics and in its links with other studies, needs to spread into the career expectations of teachers of mathematics. Mention should also be made of the retraining needs of those mathematics teachers who decide to take on responsibilities outside mathematics, in administration or pastoral work. Some of these teachers combine their new work with a continuing interest in mathematics, and their need for continuing education in mathematics, as well as in their other responsibilities, needs careful planning, so that they are involved in the development of their mathematical professional life as well as their new work.

It is sometimes suggested that, because of the continuing shortage of mathematics teachers, ways should be found of retaining them in full-time classroom work, rather than employing some of their time and energy in other tasks. It would be a long-term disservice to mathematics if, because of the shortage of mathematics teachers, able teachers were denied promotion simply because their teaching subject is mathematics. Thought needs rather to be given to ways in which they can maintain a vigorous contribution and involvement in some mathematical work, alongside their other interests.

Another phase of professional life for some teachers of mathematics is movement in and out of teaching. A spell in industry or commerce may be extremely useful to a teacher in broadening his horizons and enabling him to see the application of his work. More study needs to be given to ways of encouraging such a traffic. In just the same way that a number of teachers undertake a spell of teaching overseas or a teacher fellowship at a university, it is possible to imagine a growth of schemes designed to enable teachers of mathematics to work for a fairly extended period in industry or commerce and then to return to school with a broadened view of mathematics. Conversely, liaison between industry and education would be improved if more representatives of industry were able to work with teachers in schools, even for short periods of time.

Women teachers move in and out of teaching as the needs of their families dictate. Among these teachers there is a valuable resource of teachers of mathematics. However, while a teacher brings up a family, the situation in the schools she has temporarily left is changing; ways need to be found to help her over the difficult reorientation period as she starts work again. Her situation is in some ways comparable with that of the young teacher who is entering the profession for the first time, but her status as an experienced teacher is very different.

Problems and trends in secondary schools

The most obvious mathematical problem in British secondary schools is the continuing shortage of well-qualified teachers of mathematics. In general, schools have responded to this problem either by taking on teachers whose qualifications are inadequate for the level of work they are required to tackle, or by persuading teachers of other subjects to do some mathematics alongside their other work. This problem will soon become part of the new problem created by falling rolls in secondary schools, which will make redeployment of teachers to improve the quality of education take second place to redeployment of teachers caused by the new circumstances of shrinking schools. There is not likely to be a substantially smaller proportion of underqualified mathematics teachers in the 1980s, and means must be found of enhancing the quality of the mathematical professional life of these teachers.

Another group of problems for mathematics teachers relates to the continuing discussion about the role of the school in society. Because mathematics is increasingly used in so many areas of education and employment, it comes under the spotlight in debates about school and working life, and about the accountability of the school to the community. Teachers need to resolve questions about the balance between an emphasis on topics which industry seems to insist on, such as the imperial units still used in much of the engineering industry, and an emphasis whose rationale lies more clearly within the general education of the pupil.

A further important issue is that provided by the DES discussion document *A Framework for the School Curriculum*[18] which calls for 'an organized policy for mathematics across the curriculum, establishing the involvement of most teachers in fostering this development'. Mathematics teachers in secondary schools have traditionally thought themselves responsible for providing mathematical instruction which might be applied in other curriculum areas, but mathematics across the curriculum calls for a change of emphasis, so that teachers of different subjects work to use the contributions of all their subjects in building up the children's mathematics. In the topic and project work which is done in primary schools, opportunities for mathematical exploration often remain unnoticed, even though the same teacher is also teaching the children mathematics. Thus, the barrier between mathematics and other curriculum areas needs breaking down in primary as well as secondary schools.

In the area of curriculum development in secondary schools, mathematics teachers are beset on the one hand by vociferous criticisms from employers and the media about the lack of basic numeracy of schoolchildren, and on the other hand by the new situation created by the cheap accessibility to the public at large of electronic calculators. Calculators are only the forerunners of the micro-electronic revolution; micro-computers are beginning to appear on a substantial scale in secondary schools, and will have some effect on the mathematics curriculum, as well as producing a further diversification of the talents of mathematics teachers into computer studies. However, a substantial proportion of the curriculum effort of mathematics teachers will have to go into the development of the common system of examining at 16+. In mathematics this is likely to be more complicated than in many subjects: the same history question can be answered by many pupils at many levels, whereas a mathematics question has to be aimed much more precisely at a particular level of knowledge and ability. Devising and operating a system which is effective across a very substantial ability-range will be no easy task for mathematics teachers.

The needs of special groups of pupils are also a subject for concern. The Mathematics Supplement to the recent HMI Survey of secondary schools[19] reveals that

> A need to organise new courses for less able pupils was perceived in 68 per cent of all comprehensive schools and in 60 per cent of all secondary modern schools. In almost three-quarters of these cases the recommendation was a strong one.

Much concern has also recently been expressed about the provision for able and gifted mathematicians in secondary schools, and about the fact that substantially fewer girls than boys study mathematics in the sixth-form.

Another area of difficulty discussed in some of the contributions to this book is that of continuity of education when pupils move *en masse* from the primary or middle school to the secondary school. Mathematics, by its consecutive nature, is particularly subject to disruptions caused by lack of continuity and incorrect expectations on the part of the receiving school.

The common characteristic of all the problems listed here is that their study and solution need the *time* of teachers. Many of these problems are 'within school' problems, not to be solved only by sending teachers away on a course. They need the united work and cooperation of all the mathematics teachers in a school, and they place a particular demand on the professionalism of the head of the mathematics department. The major trend of professional life for mathematics teachers in secondary schools is likely to be a growing awareness of, and involvement in, a professional life which involves cooperative study and problem-solving with their colleagues in the mathematics department and in other departments of the school, as well as links with the community outside school. All these activities need teacher time. Teacher time is of two types: first, there is out-of-school time, in evenings, weekends, and school holidays; second, there is time during school hours when the teacher is engaged in activities other than teaching and supervising groups of children. Some professional development activities can be carried on in out-of-school time: curriculum development, discussion, and planning are examples. Other activities can only be carried on in school time when children are present: some examples are working alongside colleagues from other subject areas in 'mathematics across the curriculum', observing how pupils tackle new curriculum materials, and working alongside probationary teachers in an induction scheme. Whether professional activities take place in school or not, they have to compete with the day-to-day activities of marking and lesson preparation.

For many heads of mathematics departments in secondary schools, working with pupils is just the tip of an iceberg, the remainder being made up of organization and administration, working with colleagues, curriculum development, external examinations, links with other subjects, and the growing area of public relations with employers, higher and further education, parents, and the local community.

Problems and trends in primary schools

A difficult problem in primary school mathematics is the lack of mathematical knowledge and confidence felt by many primary school teachers. Some schools and some individual teachers work with great understanding and confidence, but the majority of teachers do not have sufficient personal mathematical grounding to feel a secure knowledge that they are doing the right thing for their pupils. This leaves them at the mercy of conflicting external pressures. The avid attendance of primary teachers at short inservice courses in mathematics is a mark of an anxious desire to understand more about mathematics and its teaching. It has become clear, however, that short courses do not provide an adequate foundation for permanent and extensive change. The improvement of the mathematics programme of a primary school is a matter not only for individuals, but for the whole staff of a school. This is particularly important as continuity of teaching is not one of the hallmarks of the British primary school – it is very usual for a child to be taught by a different teacher in each of his six years in the primary school. Even if each of these teachers is very able, the closest cooperation between teachers is necessary if a child is not to experience an annual setback as he adjusts to his new teacher's mathematical language, way of working, style of explanations, and even arrangement of calculations.

The appointment of a mathematics coordinator in each primary school has recently been recommended, and several of our contributors have discussed the tasks and problems of this teacher. A development of professional life which is to be expected in the next few years is a much more detailed working out in practice of the role of a coordinator, including his professional relationship with the headteacher, other colleagues in his own school, teachers in other schools, LEA advisers, and parents. Not all schools are large enough to appoint a coordinator to a promoted post and, in other schools, all the promoted posts may already be deployed in other ways. Which member of staff in these schools is to take responsibility for professional development of teachers of mathematics?

Some problems of children's learning to which teachers of mathematics will need to turn their attention are the relationships between the methods needed for the learning of concepts, the learning of skills, and the application of these concepts and skills in everyday situations, and in aspects of the curriculum which are not overtly mathematical. At present there is often little link between the concrete practical work in measuring and in early number work and the abstract learning of pencil-and-paper arithmetical skills which still forms much of the staple diet of many children in the later primary years. The links between the development of oral and written language and the learning of mathematics are not yet clearly acknowledged by many teachers; in particular, the role of conversation between children and with the teacher is a little-understood area of mathematics teaching. Many teachers are unclear about the levels of mathematical understanding and skill which they should expect children of different abilities to reach by the time they transfer to the secondary school.

All these problems not only need work by researchers outside the immediate classroom context, but they are also topics of immediate professional concern to which primary teachers must make their own contribution.

On the curriculum side, an issue which will need to be studied in the next few years is the role of electronic calculators in primary school mathematics. Three methods of performing 'the four rules of arithmetic' are now available: mental methods, pencil-and-paper algorithms, and by the use of a calculator. In recent years, pencil-and-paper methods have gained ascendancy, to the extent that mental methods are not always taught, and although some children work out mental methods of calculation for themselves, others do not do so, and are handicapped in their use of number in daily life. The calculator age is bound to see a reassessment of the balance of methods, and so to have an effect on the curriculum in the primary school. New skills of prediction, and estimation of the size of results, will also be needed if the calculator is to be a useful tool.

Other aspects of mathematical professional life for primary teachers are those involved in dealing with the expectations of parents and of other schools. Many parents do not understand the range of activities which children need if they are to gain mathematical concepts, and feel that the primary school is only teaching mathematics when the children are producing pages of neat sums. Not enough has yet been done to involve parents in understanding and helping their children's mathematical development. Similarly, a real two-way understanding, between primary and secondary schools, of each other's approaches to mathematics is only just beginning, and it will take time to grow, as each group of teachers builds up a professional respect and confidence in the other's way of teaching mathematics.

What are the needs and conditions for an enhancement of mathematical professional life in primary school? As in secondary schools, the first need is *time*; mathematics is only one of the curriculum areas in which all primary teachers have to work, even though it is a very important area. Some of the needed professional work can take place when the children are not present, but coordinators need time to work alongside their colleagues and with groups of children. Teachers from primary schools need time to visit and discuss with their secondary colleagues, in ways such as those described in Chapter 5. Perhaps rather more than their secondary colleagues, primary teachers need substantial in-service courses outside the school, so that they can upgrade their own mathematical knowledge and understanding, for most primary schools have fewer mathematically knowledgeable teachers already in the school than do secondary schools. However, the needs of different primary teachers are different, and ways need to be found of matching courses to the different mathematical levels of teachers, as well as of encouraging unconfident, frightened, and even antagonistic teachers to become responsible for their own mathematical development.

Priorities

In these times of zero economic growth, or of negative growth, what are the priorities in the professional life of teachers of mathematics? The first priority is not expensive. It is a recognition that the professional life of a mathematics teacher needs to be seen as a lifetime of growth and development. Mathematics itself does not stand still over a forty-year period; school mathematics can never again stand still over the forty years of a teacher's professional life. The point was vigorously made by HMI in the Mathematics Supplement of the HMI secondary survey,[19] but it applies equally to teachers in primary schools.

> In industry and in higher education mathematicians require continually to update and improve their knowledge, and people who previously needed little mathematics not infrequently have to learn to apply branches of knowledge which involve mathematical skills. There are comparable needs in school teaching. The knowledge of teachers needs to be refurbished and refreshed, and if this knowledge is to be put to the most effective use teachers need to participate in regular professional discussion of the aims which they are working to achieve.

Teachers need to participate in regular professional discussion not only of the aims; they must also study and evaluate the means by which they are trying to achieve those aims. Means may well change radically during a teacher's professional life, even if the aims remain rather more constant; and teachers constantly need to reflect on and refurbish the aims and ideals which they brought into mathematics teaching at the outset of their careers – the wish to help children to understand and enjoy mathematics, and to realize its usefulness and power. Other priorities must be ways of ensuring that teachers have time: time for thinking and for study, for planning, curriculum development, and evaluation, for reading and discussion and professional contacts outside the world of school. And the especial priority for time must be time for the team of teachers in a school to work together on the mathematical needs of their school. This demands the gradual development of the support services which are already available: LEA advisers, mathematics centres and teachers' centres, staff of universities, polytechnics, and colleges. It demands a closer integration between preservice education – which includes the study of mathematics at university or polytechnic as well as including initial teacher education – and the induction and in-service needs of the schools. It also demands the development of more classroom-oriented research into the learning of mathematics and other aspects of mathematical education.

If we fail to develop mathematics teaching into a worthwhile profession with its own professional expectations and rewards, we shall also continue to fail, as we are at present, to attract into the profession of mathematics teaching a continuing supply of able young mathematicians, on whom the future health of school mathematics depends, and for whom there are so many counter-attractions in industry, business, and other professions.

LIST OF CONTRIBUTORS

The contributions of each writer are identified by the letters in brackets after each name.

Bernard Bagnall, Dale Hall Primary School, Ipswich   [G]
John Baker, The Open University   [X]
Jim Boucher, Thames Primary School, Blackpool   [C]
Alan Broadway, Parkside Community College, Cambridge   [R]
David Brown, Archbishop Holgate's Grammar School, York   [FF]
Jennifer Caldwell, Portobello Junior School, Walsall   [X]
Dorothy Carter, London Borough of Ealing   [E]
Graham Dunn, Edge Hill College of Higher Education, Ormskirk   [BB]
David Fielker, Abbey Wood Mathematics Centre, London   [I], [EE]
David Francis, Council for National Academic Awards   [V]
Harry Fullerton, All Saints R.C. Junior School, Liverpool   [H], [W]
Mary Harris, West London Institute of Higher Education   [B]
Gwynne James, Speke Comprehensive School, Liverpool   [K]
Ron Jones, Upwood County Primary School, Huntingdon   [D]
Gillian Manning, Parkside Community College, Cambridge   [GG]

Colin Nye, Jack Hunt School, Peterborough   [J]
Barry Pinfield, Manchester LEA   [P], [S], [AA]
Sister Timothy Pinner, Christ's College, Liverpool (now at
      Homerton College, Cambridge)   [N]
Douglas Quadling, Cambridge Institute of Education   [T], [CC], [DD]
Peter Reynolds, Suffolk LEA   [O], [Y]
Hilary Shuard, Homerton College, Cambridge   [Z]
Bob Smith, Thomas Estley Community College, Broughton Astley,
      Leicestershire   [L]
Veronica Smith, King's Hedges Junior School, Cambridge   [Z]
Peter Spooner, Crossacre Junior School, Wythenshawe, Manchester [F]
Bob Stone, Stapleford Mathematics Centre, Cambridge   [Q]
John Swainston, Brookvale High School, Groby, Leicestershire   [M]
Geoffrey Wain, Centre for Science Education, University of Leeds [U]

[A] These extracts are taken from writings by:

Fima Gomberg	John Langridge
Kay Harrison	Sue Martin
Freya Horwich	Stephen Raleigh
Helen Jenkins	Crispin Taylor
Peter Kite	Jan Twigger

GLOSSARY

This glossary relates to the educational system of England and Wales, and the descriptions of terms relate only to this system. The Scottish system is broadly similar to the English system in its general outline, but there are many important differences of structure and detail, so that it would be misleading to regard the glossary as necessarily applying to Scotland. Moreover, the variety of forms of educational organization in England and Wales is such that it is impossible to describe all of them in this glossary, and exceptions will be found to many of the definitions given. The glossary is highly selective, covering only terms used in this book or essential to the understanding of contributions in it.

ADVISER. Most LEAs (q.v.) appoint one or more advisers (sometimes called inspectors or organizers) for each subject of the school curriculum. Advisers provide short courses of in-service education for the teachers in the LEA, visit schools in an advisory capacity, and help probationary teachers (q.v.).

Many advisers also have general responsibilities, such as taking part in the appointment and promotion of teachers in the LEA's service. It is rare for an LEA to have more than one mathematics adviser, and their responsibilities usually extend to primary as well as secondary schools. At present, many mathematics advisers are also responsible for the development of computer studies in schools.

A–LEVEL: *The Advanced Level examination of the General Certificate of Education.* The A-Level examination is usually taken two years after O-Level (q.v.), at 18+, by pupils who have followed a two-year academic sixth-form course. It is usually taken in three subjects in which the pupil has specialized, possibly together with general studies, which are intended to balance the pupil's curriculum.

Mathematics may be taken as either one or (less frequently) two subjects, according to the degree of specialization desired. There is a considerable variety of syllabuses, so that a pupil who takes 'single subject' mathematics may take pure mathematics alone, or in combination with mechanics or statistics, or both. Pupils who take 'double subject' mathematics often study all three branches. Mathematics may be taken in combination with any other subject or subjects, according to the timetable provision which a school is able to make. It used to be more common for mathematics to be taken in combination with science subjects, but there has recently been a growth of combinations which mix mathematics with social sciences or humanities. Passes in at least two subjects at A-Level are required for entry to degree courses, and many universities expect either three A-Levels or high grades, or both.

The A-Level examinations are administered by the same eight examination boards as the O-Level examinations.

B.Ed.: *Bachelor of Education Degree.* Until 1969, most entrants to teaching in primary schools, and many in secondary schools, followed a three-year (two-year until 1960) nongraduate course of initial training in a college of education, leading to a Certificate of Education. From 1969 onwards, the more academically able students in colleges of education were able to take a fourth year of study leading to a Bachelor of Education degree. The B.Ed. degree also became available to serving teachers as an in-service qualification. In the mid-1970s, B.Ed. degrees were restructured so that both three-year B.Ed. (Ordinary) and four-year B.Ed. (Honours) degrees became available. The Certificate of Education was phased out at the same time, so that by about 1983 all entrants to teaching will be graduates.

B.Ed. degree courses usually combine the study of education theory with that of one or two main teaching subjects, and also incorporate professional studies and teaching practice.

APU: *Assessment of Performance Unit.*   The APU was set up by the DES (q.v.) in 1975, in order to assess and monitor the achievements of school-children on a national scale, and to seek to identify areas of underachievement. At the present time, it has been decided to carry out assessment in mathematics, language, science, and the first foreign modern language, but other areas are also under consideration. A mathematics test of the eleven-year-old population was carried out for the first time in 1978, when a representative national sample of pupils was tested. Annual testing of samples of pupils at age eleven will continue, and testing will also take place at age fifteen–sixteen.

In the 1978 test, a wide range of mathematical items was used, but each pupil selected for testing was only asked to answer a small selection of the items. Pencil-and-paper testing was the major procedure, but testers also visited schools and asked pupils to carry out a variety of practical mathematical tasks. The opportunity was also taken to inquire into pupils' attitudes to mathematics, and teachers were invited to comment on the relevance of the tests used.

The report of this first set of tests was published in 1980,[20] followed by the report of the first set of mathematics tests carried out on fifteen-year-olds. The reports of the APU surveys will be presented in such a way that general trends can be identified, but not the performance of individual pupils, schools, or LEAs (q.v.).

CNAA: *Council for National Academic Awards.*   Universities in Great Britain are entitled, under their charters, to award degrees; nonuniversity institutions of higher education such as polytechnics have no such entitlement. The students of these institutions obtain their degrees through the CNAA. The CNAA operates by satisfying itself about the academic standards of an institution, its staff, and its courses; degree examinations are then set internally within the institution, and the CNAA nominates external examiners who work alongside the internal examiners. The CNAA thus validates courses provided by institutions, rather than providing its own 'external degree' examinations. CNAA committees always contain a majority of members drawn from institutions which provide similar courses.

COLLEGE OF EDUCATION: Colleges of education were formerly nonuniversity institutions which provided three-year (two-year before

1960) courses of initial training for teaching. Some also offered courses for the Postgraduate Certificate in Education (q.v.), and they also provided a fourth-year course for the B.Ed. degree (q.v.). The mid-1970s saw a major reorganization and cutback in initial teacher education in England and Wales. Many colleges were absorbed into polytechnics (q.v.). Other colleges amalgamated with one another to form colleges of higher education, which offer a variety of other courses as well as teacher education courses. A few colleges were absorbed into universities, and many smaller colleges closed.

At the same time, the Certificate of Education was replaced by three- and four-year B.Ed. degrees (q.v.). Colleges were also required to use 20% of their resources in providing in-service education for teachers; previously, their provision of, and commitment to, in-service education had varied very considerably.

CSE: *Certificate of Secondary Education.* The Certificate of Secondary Education is an examination intended for pupils of age 16+ and of average ability, somewhat below the standard of GCE O-Level (q.v.), although a Grade 1 in CSE is equivalent to a Grade C in O-Level. CSE is administered by thirteen regional examination boards, and all its committees have majorities of teachers from schools taking part in the examination. The CSE is only about fifteen years old, and has grown rapidly in popularity. In many schools up to 80% of the 16+ age-group are now entered for a public examination in mathematics, either O-Level or CSE or both.

The government has recently made proposals for a common system of examinations at 16+. This will replace both O-Level and CSE, and will avoid some of the problems of the present system, in which difficult choices often have to be made about examination entries for individual pupils. Other problems of examining a very broad band of ability will not easily be solved.

DES: *Department of Education and Science.* This is the government department responsible for the education system in England. At different periods before 1964 it was variously known as the Ministry of Education and the Board of Education. It controls the broad allocation of resources for education, but has little responsibility for detailed expenditure, most of which is controlled by LEAs (q.v.). It does not control curricula or prescribe teaching methods. The cabinet minister responsible for the DES is the Secretary of State for Education and Science. The Welsh Office is responsible for the education system in Wales, and the Scottish Office for that in Scotland.

ELEVEN–PLUS EXAMINATION.   In the selective system of secondary education, pupils were selected for grammar school education (q.v.) by means of an examination at 11+, which was organized by each LEA (q.v.). The examination usually consisted of tests of English, arithmetic, and 'intelligence' or verbal reasoning. Until its abolition when secondary schools were reorganized on comprehensive lines, the 11+ examination provided the basis for a common arithmetic syllabus for all the primary schools in a LEA; its abolition has been an important influence in making change possible in the emphasis and scope of mathematics in primary schools.

FIRST SCHOOL.   A first school is a school for children between the ages of five and eight or nine, in an area whose organizational structure contains middle schools (q.v.) for children of eight to twelve or nine to thirteen.

GRAMMAR SCHOOL.   Before the reorganization of secondary education along comprehensive lines, a grammar school was a selective secondary school which provided an academic education for pupils of eleven to eighteen.

Grammar schools usually admitted about the top 20% of the ability-range, as determined by the '11+' selection examination (q.v.). The majority of these schools have now been reorganized into comprehensive schools, but some still remain. About 5% of children attend independent (private or 'public') schools, some of which provide a curriculum similar to that of a grammar school.

HEAD OF DEPARTMENT.   In a secondary school, each subject is taught by a group of teachers who specialize in that subject. The group is led by a head of department who holds a promoted post. His responsibilities include decisions on curriculum within the subject and on entry of pupils for external examinations, choice of textbooks, ordering of equipment, and allocation of departmental staff to the teaching of particular groups of pupils. In practice, his influence can spread very widely throughout the teaching of his subject in the school and in links with other departments.

HIGH SCHOOL.   'High school' is one of the names applied to comprehensive secondary schools in some forms of educational organization. Pupils usually enter at eleven or twelve, and high schools do not usually have sixth-forms (q.v.), so that pupils then proceed to sixth-form colleges, or sometimes to upper schools.

In the earlier, selective organization of secondary education, grammar schools (q.v.) for girls were often called high schools.

HMI: *Her Majesty's Inspector.* Her Majesty's Inspectors of Schools are government employees whose chief responsibility is to keep the Secretary of State and the DES (q.v.) informed about the nation's education. In order to do this they visit schools and observe the work of pupils in the classroom. Their work also includes a variety of other tasks such as carrying out surveys of aspects of the educational system, providing in-service training, sitting on national committees, and giving advice to members of the education service on a variety of educational matters.

INFANT SCHOOL. An infant school is a school for children between the ages of five and seven. At seven, children move from the infant school to a junior school (q.v.). Although an infant school and a junior school are often on neighbouring sites, they are separate schools, and continuity in styles of mathematics teaching is a matter of negotiation between schools. Continuity is more easily assured in the case of a single primary school (q.v.), at which children remain from five to eleven.

INSTITUTE OF EDUCATION. Institutes of education were set up in 1944, usually but not invariably based on universities. Their major function was to supervise the initial training of teachers in their area, and to award the Certificate of Education to students from the colleges of education (q.v.). This function has now ceased, but many institutes of education played and continue to play an active role in the in-service education of teachers, particularly in the award of advanced diplomas. In some universities, the functions of both the institute and department of education (see UDE) are carried out within a school of education; in some cases, an institution with one of the two titles (such as the University of London Institute of Education) carries out both functions.

JUNIOR SCHOOL. A junior school is a school for children between the ages of seven and eleven. Children enter this school from an infant school, stay in it for four years, and leave it to enter a secondary school. In some areas all the children from one junior school proceed to the same secondary school; in other areas they may go on to any of a range of secondary schools.

LEA: *Local Education Authority*. There are 104 local education authorities in England and Wales, responsible for the provision of education within their boundaries. LEAs are under the political control of education committees, which are subcommittees of elected county or metropolitan councils. Within the provisions of the Education Acts, LEAs are autonomous. They provide schools, employ teachers, and are responsible for the provision of books, equipment, school meals, transport, etc.

The funding for education comes partly from local rates (property taxes), but largely from central government *via* the Rate Support Grant. The Rate Support Grant is a block grant which the local authority has discretion to share between local services such as hospitals, roads, housing, and welfare services, as well as education.

The curriculum of schools is the responsibility of LEAs, but this responsibility is largely delegated to individual schools.

MATHEMATICS COORDINATOR. In a primary school, it is usual for each class of children to be taught by a class teacher, who is responsible for all their work for a year, except possibly in specialist subjects such as music and French (if this is taught). Mathematics is almost invariably taught by the class teacher. Recently many LEAs have begun to appoint, in as many primary schools as possible, a teacher to be responsible for coordinating all the mathematical work of the school, advising colleagues, and choosing equipment and textbooks. The appointment of a mathematics coordinator is a new initiative in a field where curriculum decisions have traditionally been the responsibility of the class teacher, within any framework provided by the school.

MIDDLE SCHOOL. A middle school is a school for children of the ages of eight to twelve or nine to thirteen. Middle schools are of fairly recent creation in England – most date from the late 1960s or later – and many have yet to find a distinctive ethos of their own; they commonly regard themselves either as overgrown primary schools, or operate primary methods for the first two years and secondary methods for the last two years.

O–LEVEL: *The Ordinary Level examination of the General Certificate of Education*. The O-Level examination is taken, usually at the age of 16+, by the more academic pupils. Pupils can take it in any number and combination of subjects, and in combination with CSE (q.v.). Mathematics is second only to English language in the number of entries and is taken by nearly one-third of the age-group.

A pupil who is following a full academic secondary school course normally takes O-Level in about eight subjects.

Each of the eight examination boards provides two or three different mathematics syllabuses which have somewhat different emphases. Choice of syllabus is a matter for an individual school.

All applicants for teacher education courses who will enter teaching in 1984 or later are required to have O-Level at Grade C or higher in mathematics and English language.

OU: *The Open University*.   The Open University is a British university, entry to which is normally restricted to students over the age of twenty-one; almost all its students are part-time, and continue to work at their own jobs while studying. Much of its teaching is done by correspondence, supplemented by radio and television broadcasts. Personal tutorial support and access to computer terminals are available at local study centres.

A minimum of six full credit courses is required for the B.A. degree, and these are usually taken at the rate of one credit per year. It is also possible to take many of the OU courses as single courses without following the complete degree programme.

POLYTECHNIC.   The thirty polytechnics in England and Wales were created in about 1970, and were based on established colleges of science and technology, commerce and arts.

Most of the work of polytechnics is at degree level, and their students obtain CNAA degrees (q.v.). In the reorganization of teacher education in the 1970s, many colleges of education were amalgamated with polytechnics, so that the majority of polytechnics now have an education department.

PGCE: *Postgraduate Certificate in Education*.   Rather more than half the entrants to teaching in England and Wales now enter teaching after taking a Bachelor of Arts (B.A.) or Bachelor of Science (B.Sc.) degree in subjects other than education. Their initial training for teaching takes the form of a one-year Postgraduate Certificate in Education. These courses take place either at university departments of education (q.v.) or at colleges of education or polytechnics. Since 1973, it has been compulsory for all B.A. and B.Sc. graduates to take the PGCE before entry to teaching, except in the case of.graduates in mathematics and science, where the continuing shortage of teachers has allowed the government to continue to permit untrained graduates to enter teaching. The PGCE course lasts one academic year, about one-third of which is devoted to teaching practice in schools.

PRIMARY SCHOOL.   This is the generic term for a school which takes children of ages up to eleven. Some primary schools take children of five to eleven; in other places there are infant schools for children of five to seven, followed by junior schools for those of seven to eleven, or first schools for children up to eight or nine, followed by middle schools for those of eight to twelve or nine to thirteen.

In general, the terms primary school and primary education embrace all these structures. Children in primary schools are usually organized in classes of about thirty, and a class teacher is usually responsible for all the work of a class for a year, except perhaps in specialist subjects such as music. The children in a class are usually of the same chronological age, although 'vertical grouping' or 'family grouping' has become more popular in recent years, and 'team-teaching' is sometimes found in primary schools. However, almost all forms of organization demand that every primary teacher should teach mathematics to her class.

The mathematical background of the majority of primary teachers does not extend beyond O-Level (q.v.) and between one-third and one-half of them did not attain this standard.

PROBATIONARY TEACHER: *(Probationer)*.   On their first entry into teaching in maintained schools, teachers have to serve a probationary period of one year, on the satisfactory completion of which they are recommended to the DES as efficient.

In a small number of cases, teachers have their probation extended, and it is possible (although rare) for a teacher not to pass the probationary period, and so be unable to continue to teach in maintained schools. In 1972 the government proposed that the probationary period should become an induction year, in which the young teacher would receive specific help (both within the school and from the LEA) and would have a lighter teaching load so that he could continue his study of teaching. Because of the worsening economic climate, little has been done to implement these proposals.

SCHOOLS COUNCIL. The Schools Council for the Curriculum and Examinations is an independent body representing a partnership between teachers, the DES (q.v.), and LEAs (q.v.). It was set up in 1964, and is an important source of finance for curriculum development in England and Wales. It does not determine the curriculum, but has established many curriculum development projects, whose results and materials are made generally available to schools. Recently it has moved away from direct funding of projects to working in partnership with teachers and LEAs, so that teachers can examine aspects of the curriculum and participate directly in curriculum development.

The Schools Council also plays a coordinating role in respect to public examinations in schools; in 1970, for instance, it recommended the establishment of a single examination at 16+, which would replace O-Level and CSE.

SECONDARY MODERN SCHOOL. Secondary modern schools were a creation of the 1944 Education Act; they provided secondary education for the 80% (approximately) of the age-group who were not selected for grammar schools. When the education in an area was reorganized, grammar schools and secondary modern schools were replaced by comprehensive schools; this reorganization is still taking place in some areas.

SECONDARY SCHOOL. Secondary education is the generic term in England and Wales for the education of pupils over the age of eleven. Thus, a secondary school is a school whose pupils enter at the age of eleven or older. Pupils may enter their first secondary school at the age of eleven, twelve, or thirteen, but eleven and thirteen are the most common ages.

The majority of secondary schools are now comprehensive, but a few grammar schools and secondary modern schools remain, and in some areas reorganization is still taking place. Secondary schools whose pupils enter at thirteen are often called upper schools. Most secondary schools make provision in a sixth-form (q.v.) for those pupils who wish to stay at school after the period of compulsory schooling which ends at sixteen.

Secondary school pupils usually follow a common curriculum, which always includes mathematics, up to the age of fourteen. After this, a system of options is usually employed, but in the vast majority of schools mathematics is compulsory up to the age of sixteen. The mathematics syllabus, however, may vary considerably according to the abilities of the pupils.

SIXTH–FORM. The first five years of secondary education span the years from eleven to sixteen. Pupils who stay at school after the end of compulsory schooling (at sixteen) enter the sixth-form. Sixth-form work normally lasts for two years, and culminates in the A-Level examination (q.v.) which qualifies students for entry to higher education. Sixth-forms also usually make provision for pupils to take (or retake) O-Level subjects. For entry to some sixth-forms, at least a modest O-Level success is needed; other sixth-forms are 'open access' and, in addition to A-Level work, provide courses, often of one year, below A-Level standard.

A sixth-form is often an integral part of a comprehensive or grammar school, but in some areas, all the sixth-form work of the schools in the area is gathered together in a small number of sixth-form colleges. These cater for pupils between the ages of sixteen and eighteen, and are able to offer a greater variety of subjects at A-Level than is a school with a small sixth-form. In particular, a sixth-form college may offer several different A-Level mathematics syllabuses to cater for different interests and needs.

Among the pupils who leave school at sixteen, some follow A-Level courses in colleges of further education, which also offer a variety of technical and commercial courses.

UDE: *University Department of Education*.   Most universities in England and Wales have departments of education. The most important course provided at a UDE is the one-year course for the Postgraduate Certificate in Education (q.v.). Rather more than half the PGCE students in England and Wales are found in UDEs, including a considerable majority of the mathematical graduates. In some universities it is possible to take combined degrees in education and another subject such as mathematics, but the numbers of these students are small. Other responsibilities of UDEs include the supervision of higher degree students, usually for the M.Ed. and Ph.D. degrees, and the conduct of educational research.

University institutions with similar functions to a UDE are sometimes called university institutes of education (q.v.) or university schools of education.

UPPER SCHOOL.   A secondary school which caters for pupils in the thirteen to eighteen age-range. Pupils usually enter an upper school from middle schools, and remain there until they leave school. Upper schools normally have sixth-forms.

The term 'upper school' is also applied to the upper forms of a secondary school, the first two or three years of which are referred to as the 'lower school'.

SOME PATTERNS OF SCHOOL ORGANIZATION IN ENGLAND AND WALES

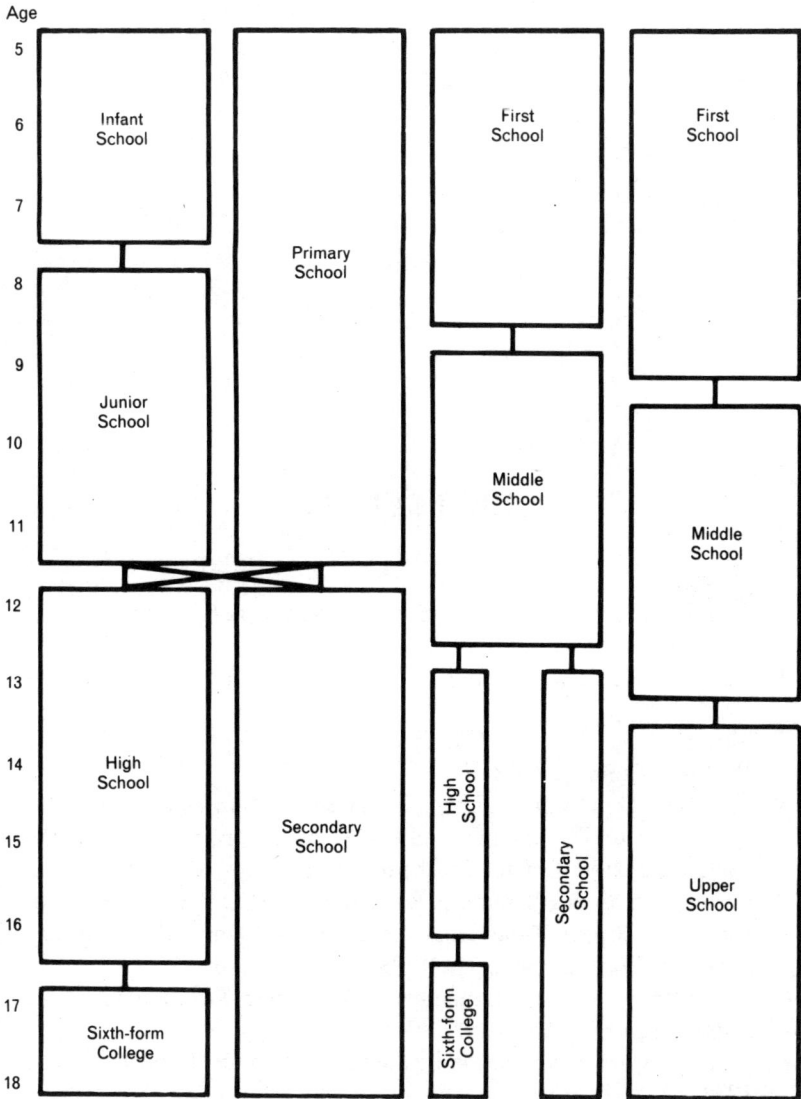

Age

Infant School	Primary School	First School	First School
Junior School		Middle School	Middle School
High School	Secondary School	High School	Upper School
Sixth-form College		Sixth-form College	Secondary School	

Compulsory schooling lasts from the age of five to sixteen

REFERENCES

1 The Royal Society, *The Training and Professional Life of Teachers of Mathematics*, Royal Society, 1976.
2 Board of Education, *Handbook of Suggestions for Teachers*, HMSO, 1937.
3 McLone, R. R., *The Training of Mathematicians*, Social Science Research Council, 1973.
4 Halls, W. D. and Humphreys, D., *European Curriculum Studies, No. 1: Mathematics*, Council for Cultural Cooperation, Strasbourg, 1968.
5 Conference Board of the Mathematical Sciences, National Advisory Committee on Mathematical Education (NACOME), *Overview and Analysis of School Mathematics, Grades K-12*, Conf. Bd. of the Math. Sciences, Washington D.C., 1975.
6 Tenth Report from the Expenditure Committee, *The Attainments of the School Leaver*, HMSO, 1977.
7 DES, *Teacher Education and Training* (the James Report), HMSO, 1972.
8 DES, *Primary Education in England: A Survey by H.M. Inspectors of Schools*, HMSO, 1978.

9 *Mathematics, Science and Modern Languages in Maintained Schools in England*, DES, 1977.

10 *The USMES Guide*, Education Development Centre, 1976.

11 Ashton, P. and Merritt, J. E., 'INSET at a Distance', *Cambridge Journal of Education, Special Issue: In-service Education*, vol. 9, nos. 2 and 3, 1979.

12 Bailey, D. E., *A Survey of Mathematics Projects Involving Education and Employment*, University of Bath, 1978.

13 *Maths after School*, North Yorkshire County Council Education Dept., 1978.

14 DES, *Aspects of Secondary Education in England: A Survey by H.M. Inspectors of Schools*, HMSO, 1979.

15 Taylor, W., *Research and Reform in Teacher Education*, NFER, 1978.

16 DES, *Education: A Framework for Expansion*, HMSO, 1972.

17 Bolam, R., et al. (eds), *1977 National Conference on Teacher Induction: Conference Papers*, University of Bristol, School of Education, 1977.

18 DES and Welsh Office, *A Framework for the School Curriculum*, DES, 1980.

19 DES, *Aspects of Secondary Education in England: Supplementary Information on Mathematics*, HMSO, 1980.

20 Assessment of Performance Unit, *Mathematical Development, Primary Survey Report No. 1*, HMSO, 1980.

INDEX

Education Act 1944, 5
eleven-plus examination, 5, 8
employment, need for mathematical
    skills,
126, 134
Evans, Norman, 90–1
examinations, external *see* Certificate of
    Secondary Education; eleven-plus;
    General Certificate of Education;
    sixteen-plus
examinations, internal, 16, 116
examiners, teachers as, 116–18
expertise, guidelines defining, 49

first schools, 7
Fletcher, H., *Mathematics for Schools*,
    35, 41

General Certificate of Education (GCE),
    5, 7, 16, 46, 116
grammar schools, 5
Great Debate, 8
group work, teachers, 78–80

HM Inspectorate, 2, 23, 27, 96–8, 129,
    135, 139
Halls, W.D. and Humphreys, D., 4
heads of department, 10, 33–4, 43,
    44–50, 75–8, 136
heads of (secondary) first year, 56-60
headteachers, 10, 27, 40
Hope, Cyril, 125

in-service training, 3, 24
    B.ed., 89–96
    compulsory, 22, 25–7
    and computer education, 111–14
    course types/programmes, 68–71,
        109–110
    criticism of, 99
    for deputy heads, 27
    follow-up, 65, 73, 113-14
    for headteachers, 27, 40
    for mathematics coordinators, 40,
        69–70
    Open University and, 96
    other than LEA, 81–114

for primary teachers, 69, 130
for probationer teachers, 25–6, 70
professional associations and,
    119–20
refresher, 26
relationship with classroom practice,
    40–2
induction of teachers, 10–11, 131–2
industry, links with, 83, 125–7, 133
infant schools, 7

James Report 1972, 10, 131
journals, 121–5
junior schools, 7

leadership skills, 29, 40
Leddy, T., 124
less-able pupils, 130–1
lesson preparation, 17, 38, 49
links between schools, 28, 51–66
local education authorities (LEAs), 6,
    67–8, 110
    advisers, 2, 18, 26, 28, 63–8

McLone, Ron, 3
Mathematical Association (MA), 11,
    119
    Diploma in Mathematical
        Education, 104–8
mathematics, as teaching subject, 17,
    18–19, 32–3, 82, 97, 128–9
mathematics across the curriculum, 25,
    82, 96, 130, 134–5
mathematics centres, 71–4
mathematics coordinators, 26, 27–39,
    40, 69–70, 136–7
middle schools, 7
Midlands Mathematics Experiment
    (MME), 6
mixed-ability grouping, 6, 83
modern mathematics, 6-7, 30
    and basic skills, 96–7
    as curriculum development, 8
    hazards of, 32
    value of new concepts, 39
monitoring *see* record-keeping

158 Teachers of Mathematics

The Harper Education Series has been designed to meet the needs of students following initial courses in teacher education at colleges and in University departments of education, as well as the interests of practising teachers.

All volumes in the series are based firmly in the practice of education and deal, in a multidisciplinary way, with practical classroom issues, school organisation and aspects of the curriculum.

Topics in the series are wide ranging, as the list of current titles indicates. In all cases the authors have set out to discuss current educational developments and show how practice is changing in the light of recent research and educational thinking. Theoretical discussions, supported by an examination of recent research and literature in the relevant fields, arise out of a consideration of classroom practice.

Care is taken to present specialist topics to the non-specialist reader in a style that is lucid and approachable. Extensive bibliographies are supplied to enable readers to pursue any given topic further.

Meriel Downey, General Editor

New titles in the Harper Education Series

Mathematics Teaching: Theory in Practice by T.H.F. Brissenden, University College of Swansea

Approaches to School Management edited by T. Bush, J. Goodey and C. Riches, Faculty of Educational Studies, The Open University

Linking Home and School: A New Review 3/ed edited by M. Craft, J. Raynor, The Open University, and Louis Cohen, Loughborough University of Technology

Control and Discipline in Schools: Perspectives and Approaches by J.W. Docking, Roehampton Institute of Higher Education

Children Learn to Measure: A Handbook for Teachers edited by J.A. Glenn, The Mathematics Education Trust

Curriculum Context edited by A.V. Kelly, Goldsmiths' College

The Primary Curriculum by A.V. Kelly and G. Blenkin, Goldsmiths' College

The Practice of Teaching by K. Martin and W. Bennett, Goldsmiths' College

Helping the Troubled Child: Interprofessional Case Studies by Stephen Murgatroyd, The Open University

Children in their Primary Schools by Henry Pluckrose, Prior Weston School

Educating the Gifted Child edited by Robert Povey, Christ Church College, Canterbury

Educational Technology in Curriculum Development 2/e by Derek Rowntree, The Open University

The Harper International Dictionary of Education by Derek Rowntree, The Open University

Education and Equality edited by David Rubinstein, University of Hull

Clever Children in Comprehensive Schools by Auriol Stevens, Education Correspondent, The Observer

Values and Evaluation in Education edited by R. Straughan and J. Wrigley, University of Reading

Middle Schools: Origins, Ideology and Practice edited by L. Tickle and A. Hargreaves, Middle Schools Research Group